James Miller

Reference Book of Ireland

Contains a complete list of provinces, counties, baronies, cities &c.

James Miller

Reference Book of Ireland

Contains a complete list of provinces, counties, baronies, cities &c.

ISBN/EAN: 9783337325497

Printed in Europe, USA, Canada, Australia, Japan

Cover: Foto ©ninafisch / pixelio.de

More available books at **www.hansebooks.com**

REFERENCE BOOK

OF

IRELAND.

CONTAINS A COMPLETE LIST OF
PROVINCES COUNTIES, BARONIES, CITIES, PARISHES AND VILLAGES,
WITH THEIR LOCATION, POPULATION, &c., &c.

COMPLETE MAP OF IRELAND, SHOWING RAILROADS, POST ROADS, &c.

ILLUSTRATED.

COMPILED FROM LATEST STATISTICS, AUTHENTIC INFORMATION, &c., BY
JAMES MILLER.

NEW YORK:
COOKE & COBB, PUBLISHERS,
30 & 32 WHITEHALL STREET.

1877.

Entered according to Act of Congress, in the year 1877, by JAMES MILLER, in the office of the Librarian of Congress at Washington.

INTRODUCTION.

Ireland is a detached island of an oblong form; is situated between Latitude 51°, 26′ and 55°, 21′ North, and Longitude 5°, 20′ and 10°, 26′ West; separated from England and Scotland by the Irish Sea, and projects further West into the Atlantic Ocean than any other portion of Europe.

The length from Northeast to Southwest (which is its greatest distance) exceeds 300 miles. The average breadth is about 180 miles, and in several places less.

It partakes largely of England's genial nature, and requires but common industry to make it extremely productive in all the comforts of life.

Its estimated extent is about 40,000 square miles, and its shores show a continuous waved outline on the East coast, but deeply indented by numerous inlets on its other sides, with rock bound coasts North and West. Surface, mostly level or undulating. Most of its mountains, as the Wicklow and Mourne mountains, and those of Galway, Donegal, Mayo and Kerry, are in isolated groups towards the coasts and extremities of the Island, surrounding a central plain of Limestone formation, comprising a large extent of bog land, and traversed only by a few low hill ranges, and the Sliebhloom mountains. McGillicuddy Reeks, in Kerry, of which Carran Tual is the highest summit in Ireland, rises to 3,414 feet above the sea. Granite, flanked by Silurian Strata forms the Wicklow range of mountains on the East coast. The same primary Strata prevails in the Mourne mountains, and an extensive trap formation in Antrim of 800 square miles is succeeded by Clay and Slate on the West. Granite also appears in the Galway mountains, and Red Sandstone extensively prevails in Munster. The plain in the centre is formed of mountain Limestone and contains 7 Coal districts; the Leinster, or Castlecomer, the Sleeve Aada in Tipperary, the Munster, Loch Allan, Monaghan and Tyrone, besides a small Coal field in County Antrim. The coal found South of Dublin is Anthracite; that in the North is Bituminous. Copper and Lead are found in the Silurian and mountain Limestone. Chalk beds and Green Sand formations partially prevail, but no tertiary beds.

The principal rivers are the Shannon, Suir, Liffey, Barrow, Blackwater, Nore, Bamfoyle, Boyne, Bandon, Slaney and Erne. The principal Lakes are Loughs, Neagh and Erne in Ulster, Loughs, Allen Corrib, Mask and Deirgeash, in Connaught, and the famed Loughs of Killarney in Munster, near the Southwest extremity of the island. The aggregate surface of these Loughs is estimated at 336 square miles; Lough Strangford or Belfast Lough, Lough Foyle and Lough Swilly on the North and Northeast coasts are inlets of the sea. other principal inlets are Carlingford, Dundrum, Dundalk and Dublin Bays on the East, and Bantry, Dunmanus, Tralee, Dingle, Galway, Clew, Blacksod, Killala, Sligo and Donegal Bays, with the estuaries of the Shannon and Kenmare rivers on the West coast. Ireland is much indented by the sea, no locality being above 50 miles from the shore, and this conduces to the general mildness of its climate. Mean Summer temperature, 60° 6, mean Winter, 39° 9; mean temperature of year, 49° 6. Annual fall of rain, 30 to 40 inches; the greatest fall being in the South and Southwest. The Arbutus and the broad leaved Myrtle grows luxuriantly in some of the Southern Counties, and other plants of South Europe flourish, while Alpin plants of the extreme North of Europe are also found in some of the mountains. Nine-tenths of the lands were confiscated by the government of Cromwell and William the III. and bestowed on English proprietors, by whose descendants the bulk of Irish property is still possessed.

The wool produced in Ireland has been estimated to amount in annual value to £400,000 Sterling. Woolen stuffs are made in centre of the Country, besides mixed stuffs and silks in Dublin. Waterford and other principal cities. The conversion of Grain into Flour and Meal has become an extensive business. The principal exports are raw products, such as Wheat, Oats, Flour, Butter, Bacon, Beef, Eggs, Wool, Flax, Linen, Ore and Spirits ; Imports comprise Coal, Culm, Fish, Salt, Woolen and Cotton Yarn, Fabrics and colonial produce. Value of exports amount to about £22,000,000 Sterling Annually ; of imports about £18,000,000 Sterling, mostly to and from Liverpool, Bristol, Glasgow and other British ports. The principal commercial ports are Dublin, Belfast, Limerick, Cork, Waterford and Londonderry. The Grand and Royal Canals with Railroads intersect Ireland throughout from East to West, and from North to South. The population of Ireland in 1871 was 5,412,377, distributed as follows : Leinster, 1,457,635, Munster, 1,393,485, Ulster, 1,833,228, Connaught, 846,213. The general valuation of houses and lands in Ireland in 1871 was £13,257,673. There were in the Workhouses 48,926 persons, in Hospitals 2,894, and in public Lunatic Asylums 7,116 ; there are 6,742 idiots, 9,763 lunatics, 81,000 paupers, 3,467 convicted prisoners, of persons 100 years old and upwards there were in Leinster 27 men and 62 women, in Munster, 101 men and 187 women, in Ulster 92 men and 133 women, and in Connaught 39 men and 83 women.

The population is very uncertain, in consequence of the large immigration, together with various other causes, but at the present time the population may be estimated at 6,000,000—the census of 1871 giving 5,412,377. Ireland having less barren land than either Scotland or England has its advantage in being capable of abundant supplies to England. It is divided into four provinces, each province containing several Counties. The Parliament is now incorporated with the British, but they retain **all their Law Courts and other establishments for the distribution of Justice.**

Ireland has since **the year 1800 been represented in the British House of Lords by twenty eight temporal and four spiritual peers ;** (elected for Life,) **who take office in** rotation, and **since the Reform Act in the House of Commons, by 105** members **; two elected by each County, two by each of the cities, Dublin, Cork, Limerick, Belfast, Galway and Waterford, also Trinity College, Dublin ; and by each of twenty-five other boroughs.**

The local government is vested **in a** Lord Lieutenant assisted **by the Secretary for Ireland and a Privy Council** nominated by the Crown, besides an indefinite **number ; the Bishop of** Meath being always one, *ex-officio*. **The** Judicial power is with the Lord Chancellor, the Master **of the Rolls** and twelve Circuit Judges.

In 1834 the religious denomination were, Roman Catholics 6,427, 712 and Protestants **of all Sects** 1,500,220. **The Protestants are mostly confined to** Ulster **and portions of Leinster The** Protestant Church has **two** Archbishoprics **(Dublin** and Armagh) and ten Bishoprics. **The Roman** Catholic **Church (**unendowed**) has** four Archbishoprics **and twenty five** Bishoprics.

The most important institutions for instruction are **the University of Dublin,** Trinity **College, the Queen's** Colleges **of** Belfast, **Galway and Cork, in connection with the Queen's** University in Ireland, (Established in 1850,) **the Roman** Catholic College of St. Patrick, Maynooth College, **St. Jarlaths,** Tuam, **Carlow,** Armagh and Thurles, the Belfast Academical **Institution, St. Kyran's,** Kilkenny, Clongrowes, County Clare, All-Hallows, **Drumcondra,** near Dublin, Missionary College, endowed Mercantile School, endowed Classical Schools. Parochial Schools, National Educational **Schools,** Church educational, Kildare Place **Schools,** Christian Brother **Schools** and Sunday Schools. The National Schools, aided by Parliament grants in 1851 of £104,577, **amounted in** 1862 **to** 6,576, attended by 978,326 children.

Christianity was first introduced into Ireland in the year 432 by St. Patrick ; from the eighth to the twelfth century the Country was a continual scene of warfare between the various Kings and their chiefs, and in the year 1174 Henry the VII. of England conquered the country, which he portioned out among his Anglo-Norman followers.

During the reign of King John the division into Counties took place, and English laws and customs were then partially introduced, still the Irish continued to resist the government of Henry the VII. James the I. and Elizabeth, by repeated rebellion and outbreaks ; the most important of these were in 1641, 1698, 1798 and 1848.

In the year 1800 a union with Ireland and England took place ; in 1829 the Roman Catholic Emancipation Act was passed ; in 1832 the Irish Reform Bill; in 1838 the Poor Law Bill; in 1844 the Irish repeal agitation was at its height ; in 1847 the great famine took place in consequence of a failure in the potato crop ; in 1848 an insurrection of the people took place but was subdued ; in 1868 the disendowment of the established church took place ; in 1870 the Land Bill was passed ; and in 1871 the Party Possession Act was passed.

The great decrease of the population of Ireland consequent with the calamities of 1845, 1846 and 1847, has effected more than anything in showing the result of those calamities before the mind forcibly.

Taking periods of ten years, the Commissioners of Census gives us the following numbers, viz : 1811, 5,937,856 ; 1821, 6,801,827 ; 1831, 7,767,401 ; 1841, 8,175,124 ; 1851, 6,661,830 ; 1861, 5,757,821 ; 1871, 5,412,377 ; an increase of population until 1841 to 1845. Three years of a very meagre crop, (1845, 1846, 1847,) together with a very large Immigration is the cause of this large decrease in population. It is estimated that during the period from 1841 to 1851 about 1,300,000 immigrated from Ireland, or about 130,000 Annually.

Population of Ireland at each Decennial Census, from 1821 to 1871

	MALES.	FEMALES.	TOTAL.	PER CENT.
1821	3,341,926	3,459,901	6,801,827	
1831	3,794,880	3,972,521	7,767,401	14.19 increase.
1841	4,019,576	4,155,548	8,175,124	5.25 "
1851	3,212,523	3,361,755	6,574,278	19.58 decrease
1861	2,839,370	2,961,597	5,798,967	11.79 "
1871	2,638,741	2,773,636	5,412,377	6.72 "

This table shows a steady decrease since 1841.

Value of Land and Buildings is between £12,000,000 and £15,000,000 Sterling; number of vessels employed in the fisheries, about 14,000, manned by about 59,000 persons around the coasts where Herrings, Pilchards, Cod, Ling and Hake are among the most plentiful fish taken and in the estuaries where Salmon and Eels are abundant. But the salt fish consumed in the country is still imported, chiefly from Scotland and other places. Mineral products comprise Marble of the finest quality, Coal, Copper, Lead, Antimony, Manganese, Fullers Earth, Slate, and Peat from the bogs, which forms the principal fuel for fire and is of high importance owing to the general deficiency of timber. Manufactures consist of Paper, Glass, Tobacco, and especially Linen Goods, the chief seat of latter is Ulster. The Malt trade and distilling of Whiskey are extensively carried on.

The Province of Munster has 9 territorial Divisions, 6 Counties and 3 Cities.

The Province of Ulster has 10 territorial Divisions, 9 Counties and Carrickfergus.

The Province of Connaught has 6 territorial Divisions, 5 Counties and one town.

The Province of Leinster has 15 territorial Divisions, 12 Counties, one town and two Cities.

Territorial Divisions and Extent of each province and County

PROVINCE OF MUNSTER.

Counties.	Baronies.	Parishes.	Towns & Villages.	Total Area. Acres.
Clare..	11	80	728	829,934
Cork	23	251	6,515	1,846,333
Kerry	8	87	807	1,186,126
Limerick	13	131	2,759	680,842
Tipperary	12	193	2,359	1,061,731
Waterford	8	82	1,525	461,553
Total	75	824	14,693	6,064,579

PROVINCE OF ULSTER.

Counties.	Baronies.	Parishes.	Towns & Villages.	Total Area. Acres.
Antrim	15	75	1,908	745,177
Armagh	8	28	778	328,076
Cavan	8	36	502	477,360
Donegal	6	51	479	1,193,443
Down	10	70	2,211	612,495
Fermanagh	8	23	210	457,195
Londonderry	6	43	1,559	518,595
Monaghan	5	23	304	319,757
Tyrone	4	42	710	806,640
Total	70	390	8,790	5,475,458

Territorial Divisions and **Extent of each Province and County.**

PROVINCE OF CONNAUGHT.

Counties.	Baronies.	Parishes.	Towns & Villages.	Total Area. Acres.
Galway	18	120	1,801	1,566,354
Leitrim	5	17	451	392,363
Mayo	9	73	848	363,882
Roscommon	9	58	768	607,691
Sligo	6	41	460	461,753
Total	47	309	4,327	4,392,043

PROVINCE OF LEINSTER.

Counties.	Baronies.	Parishes.	*Towns & Villages.	Total Area. Acres.
Carlow	7	47	602	221,342
Dublin	10	99	1,820	222,714
Kildare	14	116	490	418,436
Kilkenny	11	140	628	**508,811**
Kings	12	51	902	493,985
Longford	6	26	364	269,409
Louth	6	64	728	201,434
Meath	18	146	464	579,899
Queens	11	53	1,117	421,854
Westmeath	12	63	628	453,468
Wexford	9	144	2,392	576,588
Wicklow	8	59	341	500,178
Total	124	1,008	1,476	4,871,118

COLONIZATION
OF
IRELAND.

WESSELING the latest editor of DIODORUS, acknowledges he cannot account for Ireland being thus named instead of *Ierne, Iouernia,* and *Iernis*. But DIODORUS who had penetrated far into the North of Europe, there first heard and has happily preserved the genuine name of our Isle, a name almost two thousand years old, and yet unaccountably passed over by all our antiquaries; a name which removes every difficulty about the country designed by DIODORUS.

Iri, or as now written *Eri* in Irish, is the great Isle. In Teutonic *Er-aii,* contracted into *Eri,* is the farther Isle. It received this appellation from the Teutonic tribes, who then possessed Europe, and has been invariably used by them in every age. Here are the proofs:

A. D.
540. GILDAS left the school of ILTUTUS in *Wales* and went to *Iris*.
870. In Islands Landnamaboc, one of the oldest Islandic Sages, *Ireland* is named *Ir-land*. In King Alfred's Anglo Saxon translation of *Orosius,* Ireland is styled Ireland.
891. *Three Irishmen,* says the Anglo Saxon *Chronicle,* came in a boat from *Yr-land*.
918. In the same record under this year our Isle has the same name:
1048. In the same Chronicle, HAROLD flies to *Yr-land*.
1105. ELNOTH in his life of St. Canute calls the Irish *Iros*.
1141. ODERICUS VITALIS styles the Irish *Irenses,* and their country *Ire-land*.

In Woramus's Runic Literature, the Irish alphabet is called *Iraletur* The identity of Diodorus' *Iris* with the *Iris, Ira, Iros, Ireuses, Ire* and *Ir* of the Gothic and Teutonic people, and that traced for above six hundred years clearly evinces that this Greek Author has preserved the genuine and original name of our Isle. There are other proofs no doubt which have escaped the writers research as to the change of *Iris* into *Ierne*, whoever is acquainted with the alteration of words by Greek dialects and the effect of their epenthesis and paragogue will easily account for the mutation.

If it be asked why this original name has been hitherto unnoticed, the answer seems to be, that antiquaries find it much easier to build systems on conjectures than to laboriously enquire after truth and certainty.

The earliest notice in Roman writers of the name of Ireland (Hibernia) is in Julius Cæsar and was given probably by him or his countrymen from its supposed coldness, for it was the practice of antiquity to give appellations to countries and people from their situation, productions, or some peculiarity. Strabo, who wrote long after Cæsar, describes Britain as frigid from its vicinity to the North, and Ireland as scarcely habitable from its coldness. Hibernia was then an appellation suitable to such conceptions.

That the Romans had separate Maps of their whole Empire, and even of parts not under their dominion (as was the case of Ireland) has been shown. Whether Balbus's Commentary containing the names of cities, rivers, promontories and tribes was ever published or at what time we are nowhere told. It is certain Marines of Tyre, and Ptolemy the celebrated astronomer and geographer of Pelusium, obtained information of these and transmitted it to posterity. Ptolemy flourished A. D. 150; it might therefore be expected that the names of places in Ireland which he records would have been purely Celtic; this our native antiquaries positively deny, unless in a few instances. The country, particularly the maritime parts, was possessed at different times by such various tribes of foreigners, that we need not wonder at the instability and change of names in those distant ages. The Celts, however were the majority and preserved their language. They adopted the religion and manners of these foreigners, causing a mixed superstition of Celtic and Scythic to spring up, which both British and Irish writers call, (but very improperly,) Druidic; for the Druids were the Priests of the Celtes. On this distinction, and on this alone, rests the true and accurate explanation of the antiquities of Ireland. To establish this point it will therefore be necessary to detail with some minuteness the names and progress of the foreign colonies which arrived here.

CAMDEN is explicit that Ireland was originally peopled by Britons, but after, (from the revolutions arising in other countries,) Gauls, Germans and Spaniards were compelled to seek refuge here; SPENSER, who published his "View of Ireland," a few years after CAMDEN, tells us the Gauls were the first inhabitants of Britain and Ireland: that Gauls from Belgium and Celtica settled in the South, Scythians in the North, and Spaniards in the West of Ireland; as to the latter, he doubts whether they were Gauls or of some other country. He is correct in making the Gauls or Celts the primeval possessors of Britain and Ireland. But not so when he says the Gauls from Belgium were the same people. In the infancy of antiquarian disquisitions such errors are pardonable. He confirms the Scythic derivation of the Irish by an ample comparison of their customs and manners.

CÆSAR informs us that Gaul (or now France) was divided between three races of men, the Celts, Belgæ and Aquitani, who differed in language, manners and laws. He confounds the Celtic and Belgic practices, calling them Druidic, and in this he has been but too closely followed by subsequent writers. The Celts having colonized Britain passed from thence into Ireland. Hear what a man of consummate abilities advances on this subject: without recurring, says he, to the authority of story, but rather diligently observing the law and course of nature, I conjecture that whatever is fabled of the Phœnicians, Scythians, Biscayners, etc., of their first inhabiting Ireland, that the places nearest Carrickfergus were first peopled, and that by those who came from the parts of Scotland opposite thereto. He thinks the Britons might come from Holyhead or St. David's Head but that the primitive possessors arrived from Scotland, the passage being short and easily performed in the frailest boats. The almost identity of the *Erse* and *Irish* is complete evidence of the fact. The Irish are not descended from the Welsh Britons, because their dialect greatly deviates from the Irish; insomuch that LHUYD, LEIBNITZ and ROWLANDS acknowledge the Welsh to be but a secondary colony; being Cimbri, Cumri or German Celts. The original Irish were then Celts, who about 300 years before our era were disturbed by the Fir-bolgs, or Belgæ, a branch of the great Scythian Swarm.

The Irish Fir-bolgs were Belgic men, Viri Belgici, or Belgæ, from the Northern coast of Gaul. They possessed no inconsiderable portion of Britain before the arrival of the Romans, and by Richard of Cirencester, are said to have come here a little before Cæsar's attempt on Britain. Ptolemy mentions the Menapii and Cauci in Ireland in the middle of the Second Century; they must have come from Belgic Gaul and Germany, for we meet with no trace of them in Britain; Menapia in Wales being founded by the Irish Menapii. This Teutonic people inhabited the sea coast of Wexford and Waterford, and by the Irish are called Garmans or Germans. Our antiquaries assure us these Belgic tribes divided Ireland into five provinces, and particularly held Connaught and gave it Kings to the end of the third Century. Numberless places were called after them, and many families are derived from them; as the O'Beunachan's of Sligo, the O'Layns of Hymania; the Nials, McLaughlins and others are of Scandinavian ancestors. LHUYD puts an end to all doubt as to the power of the Belgæ in this Isle, by exhibiting a long list of words springing from the Teutonic and by adding:—We have no room for supposing, unless it be in a very few examples, that the Irish have borrowed these words from the English; because they are extant in the old Irish MSS. written before the union of the two nations; and moreover, they have several hundred Teutonic words that are not at all in the English.

The Picts, another Gothic or Teutonic people, early established themselves here as they had long before in Scotland. The same may be said of the Scots, both were Scythians and part of the Saxon nation; which, in the middle of the fifth Century, as we learn from Stephen of Byzantium was seated on the Cimbric Chersonese. Part of this people settled in Norway, and from thence sent colonies to Scotland, where they were called Albin Scutes; some came to Ireland and were named Irin Scutes. Hence SIDONIUS APPOLLINARIS in the fifth Century, speaks of them as a kindred people, who united in pillaging the Roman provinces.

It is conjectured, that the Scots came to our Isle two or three Centuries before the Nativity, and as to their name that seems not derived from a city or particular place, or ferocity or eminence in war, but from their original country, Scythia. USHER has shown that they were distinguished by this appellation from the third to the twelfth Century, and of course were the dominant people. After the settlement of the Balgæ, Picts and Scots in Ireland, every gale wafted over innumerable hordes of Northern rovers, these the Irish called Fomora, from Fomoire, or Finnland. There is an isle in the Baltic on the coast of Holstein named Femera or Femeren, where probably some powerful piratical chief reigned, who united under his command Danish, Swedish, Finnish, Iutish and Norwegian adventurers in predatory voyages, as was common in the middle ages, and which the words of O'Flaherty seem to intimate. In the age of Tacitus, the Finns were mere savages; afterwards Finnland contained six provinces and various tribes, and became, as we see, superior to their neigh-. bors. O'Flaherty relates, that Tuathal, an Irish prince, married Bania the daughter of Scalius, King of Finnland, about A. D. 130. O'Brien remarks that Tuathal, after changed into Tothil, Tohill and

Toole, was the same as Totila among the Goths, and that many Gothic names are to be found among the Scots or Irish; and O'Conor, from this marriage, concludes, that a close intercourse was maintained with the nations bordering on the Baltic in the second Century. However, the reigns of Tuathal and Scalius are ante-dated by some ages. So famous and respected were the Finns in Ireland that the word Finn was used as an honorable addition to the names of princes, as Fiatach-Finn, Fiah Finnoladh, Finn-Nachta, etc. The districts seized by the Finns were named Bescha-na-Fene, and their monstrous stone monuments, Leabthachana Bhfeinne, the beds or tombs of the Fene or Finns. They had the Bhearla-na-Fene or Finnish dialect.

The Irish intermarrying with them, formed a militia to protect the coast against their marauding countrymen. This was the body of National forces, celebrated in Irish romantic history under the name of Fionn Eirionn, and led by the great Fin MacCumhal, their general. It is pretended that the names of the stations and officers of this Finnish militia are still extant. Among the latter we find Oshen MacFinn, Fian MacFenrasse, Boge MacFinn, Row MacFinn, and Rogsklaygh MacFinn. Camden informs us that the tales and songs concerning the giants Finn MacHuyle and Oshin MacOwen were popular among the Irish in his time. Let every reader appreciate the value of these traditions and also those in Mr. Macpherson's Ossian.

Another colony of Northerns are recognized by our historians under the names of Tuatha de Danans, Danir, Dansfir or Danes; they came from Denmark. O'Flaherty saw no objection to this, but, that the name of Dane was not known until the sixth Century, however, as he well observes, like those of the Picts and Scots, the name might have been long known among the people of the North before the Romans became acquainted with it; O'Flaherty allows they spoke the German or Teutonic and inhabited the cities Falм, Goria, Finnia and Maria in the North of Germany. In the black book of Christ Church, Dublin, the arrival of the Danes here, before the age of St. Patrick is recorded. O'Conor with the scantiness of information asserts that the Scandinavian tribes which infested Ireland were not distinguished by particular names. Bishop Nicholson said that he once designed to give the Easterlings or Ostmen a Chapter, but summing up the evidence he found they did not deserve such regard, but from the specimen here given, it will be seen that the labor of investigation and not materials were wanting.

The Leathmannice or Lettmanni were another tribe settled here. They came from Letten, Letitia, or Lettenland, a part of Livonia. The name of the river running through Dublin is, in GIRALDUS CAMBRENSIS, *Avon Liff*; in old records, Avene Liff, and Avon Liffy. CAMDEN will have it to be the Libnius of Ptolemy, but Libnius is the bay of Sligo. Avon Leivi or Lifi is the river of the Leivi, a tribe adjoining the Lettmanni. Dublin, in Ptolemy, is called Eblana; a true Teutonic name from Eb-land, as the sea at ebb-tide left uncovered a strand of eight thousand acres; a sight very striking to the northerns. Dublin seems to have derived its name from Duflin, a town in Scandinavia; its Irish name of Baileacleath, or the town of hurdles, is a hybrid compound of Irish and Gallic.

The Martinei were a Belgic tribe probably from Martiniana in Zealand. O'Flaherty calls them the remains of the Belgæ.

Whether the Ostmanni or Ostmen, who ruled in Ireland, were a particular tribe or a general name, has been doubted. SNORRO says they came from Sweden. NICOLSON and GIRALDUS bring them from Norway. MURRAY denies their ever having left the Baltic, or invaded Ireland. This assertion is no proof, especially as remote tribes at this time came to Ireland. There were other Northern tribes; as the Gottiac, the Gaill and others arrived and obtained settlements here.

This system of Northern colonization has been supported by domestic and other writers, and though these various tribes spoke a language radically the same, yet they had different dialects which are distinctly noticed. Thus the Belgæ used the Belgaid or Teutonic; the Fene, the Bhearla-na-Fene; the Gaill, the Gaoileag; the Saxons, the Sagsbhearla; the Scots, the Scotbhearla; and there seems to have been a common language, made up of all, like the lingua Franca, and named Bhearla-na-Teibidh. Hence the Celtic became the most corrupt of any living language, and it is fortunate that it was not totally annihilated. Dr. O'BRIEN explains why it was not. "The Northern rovers, he says, always came in small bodies, and when landed were usually employed by one party of the natives against the other, by thus weakening both they were better able to establish themselves; besides they carried no women in these expeditions, but procured wives from the natives, whereby they and their children insensibly lost their native language." And this he exemplifies in the case of the English who came over in small parties after the conquest of Ireland.

In a word without indulging any idle or absurd hypothesis or conjecture, but taking the evidence of ancient writers as they fairly lie before us, there are ample grounds for believing the first Northern invasion of this Isle was many ages before the incarnation. Ptolemy proves the existence of German tribes here in the second Century, and Latin poets and historians evince the connection between the Irish and Northerns to the end of the fourth Century. The author of the Eulogium particularly remarks the invitation of the Gothic nation of the Picts into Britain, by Gratian and Valentinian, A. D., 382. Some time after Gratianus Municeps drove the Irish back to their country; but on his death in A. D. 407 they returned and brought with them the Scots, Norwegians and Dacians or Danes and wasted Britain. Throughout the fifth Century they infested England, and about A. D. 450, the Anglo Saxons arrived in that Island. The perpetual wars excited by these foreigners was as subversive of literary repose as it was destructive of literary memorials, and Irish writers unanimously complain of the latter being lost in these convulsions. Very little can therefore be expected previous to the ninth Century; from that time the

Northerns themselves had some imperfect records of their achievements, and partly supply the defects of our domestic annalists. Thus the Icelandic chronicles have the names of Glromal in A. D. 890; of Murchard, about A. D. 962; of Conchobar in A. D. 1018, and of Dubnial and Kyriawal in the ninth Century, as Kings of Ireland. But the most extraordinary omission of our historians is their not enquiring who Turges was, from whom descended, and when he flourished. Instead of these interesting facts we are amused with childish tales of his cruelty and amorous adventures. It would have been strange indeed of a leader who subdued the Irish, castellated and garrisoned their country, and with a triumphant army for many years held sovereign sway, should not have found one Scald to transmit his name and actions to posterity. The Icelandic records introduce us to this celebrated chief, under the name of Thorgils. The Irish not using the letter "H," but as an aspirate, and dropping one where two consonants come together, make from Thorgil's Torgis, thus Torges or Torgesius, an Ostman, was bishop of Limerick, and Thorgils is, at present, in Norway pronounced Torges.

Harald Harfagre was monarch of Norway about A. D. 890; he gave to Thorgils and Frothe, (two of his sons,) a well appointed fleet to plunder the coast of Scotland, Wales and Ireland. They landed in Dublin and reduced it under their power; Frotho was taken off by poison, but Thorgis reigned long in that city, and at length fell by the machinations of the Irish. Thorgils was attended by 120 ships and numerous forces; the Northerns dispersed over our Isle, quickly flocked to his standard and recognized the son of the great Harald; for thirty years he possessed the sovereignty of Ireland. He built castles, forts and wards, cast up trenches, banks and ditches for safeguard and refuge, was enamoured with the fair daughter of O'Melaghlin, King of Meath, who agreeing to send his daughter to him, accompanied her with sixteen young men in female attire, who dispatched Thorgils with their skenes—such is the Irish account.

The interval between the Irish and Icelandic accounts is probably not very great.

Such is the scheme of colonization concisely sketched out and now laid before the reader. It admits of enlargement even to lassitude. This scheme is founded on the sure basis of written authorities, and which, while it dispels the obscurity, casts a steady light on every branch of Irish antiquities.

ILLUSTRATIONS.

ABBEY AND CHURCH OF AGHABOE.

Aghaboe is situated in the Principalities of Ossory, which included the whole County of Kilkenny, called Lower Ossory, and a greater part of the County Queens, named upper Ossory, being co-extensive with the Bishop's jurisdiction at this day. Aghaboe Abbey was founded at end of the sixth Century. (See parish of Aghaboe.)

A NORTHWEST VIEW OF THE ABBEY OF ATHASSEL.

Athassel was founded by William Fitz Adelm de Burke about the year 1200, in the village of Athassel, three miles from Cashel, for Cannons Regular of the Order of St. Augustine. The ruins of this Priory speak its former magnitude and splendor. The choir is forty-four feet by twenty-six.

A SOUTHEAST VIEW OF THE ROCK OF CASHEL.

Malachy O'Morgair, about 1135, erected at the Abbey of Saul, two stone roofed Ozyts, seven feet high, six long, and two and a half wide, with a small window at one side. But that of the greatest magnitude and best architecture is Cormac's Chapel at Cashel.

CHAPEL AT HOLYCROSS.

This Abbey is situated in the County of Tipperary, about two miles from Thurles Donagh. Carbragh O'Brien, King of Limerick, founded it in 1169 in honor of the Holy Cross, Saint Mary and Saint Benedict, for Monks of the Cistertian Order.

OLD LEIGHLIN.

Leighlin, or rather Lethglen, the half enclosed valley, is situated in the barony of Idrone and County Carlow, in a recess of the Slieumargah mountains. Monastic legends ascribe the foundation of the Church and Episcopal See of Leighlin to Saint Laferian, about 632.

PLACES IN IRELAND:

ALPHABETICALLY ARRANGED.

ACHILL; or EAGLE ISLAND. An island off the West Coast of Ireland, County Mayo. Circumference, about 30 miles. Area, 35,283 acres. Population of parish, 6,392, mostly occupied in fishing. At its Northeast end is a Protestant Missionary establishment. Its west point forms Achill Head 2,222 feet in the elevation latitude 53° 59′ North. Longitude, 10° 12′ West. Achill Beg is an island immediately South the foregoing.

ACHONRY.—A parish of County Sligo, 15 miles South Southwest of Sligo. Area, 69,896 acres. Population, 17,986.

ADAMSTOWN.—A village and parish of Wexford County, and 17½ miles West Northwest Wexford. Area of parish, 8,134 acres. Population, 2,037. Here the barn of Scullabogue was burned during the rebellion of 1798, with a number of persons inside.

ADARE.—A decayed town and parish, County Limerick, on the Maig, 10 miles Southwest Limerick. Area of parish, 12,093 acres. Population of parish, 4,902; of town, 1,095. It has a long stone bridge, an old castle and some monastic remains.

AGHABOE.—A parish in Leinster, Queens County, 11 miles Southwest Maryboro. Area, 18,702 acres. Population, 6,310. Formerly a Culdee establishment, and in early times the ecclesiastical metropolis of the Ossory Territory.

AGHADOE.—A parish of County Kerry, 27 miles Southwest Castlemaine. Area, 19,888 acres, including 1,200 of water. Population, 4,897. The ruins of an ancient castle and the cathedral still remains.

AGHRIM; or AUGHRIM.—A parish of County Galway 13 miles Northeast Loughrea. It is famous in British history for the great victory obtained here in 1691 by the troops of William III. over those of James II. Area, 7,252 acres. Population, 2,127.

AHASCRAGH.—A town and parish of County Galway, on the Ahascragh, 17 miles North Northeast Loughrea. Area, 17,305 acres. Population of parish, 5,380; of town, 775. The town is neat and clean.

AHOGILL.—A parish of County Antrim, Ulster, 3 miles Southwest Ballymena. Area, 32,987 acres. Population 23,622.

ALLEN.—(Bog of).—Is a collective term applied to the bogs East of the Shannon in Kings County and Kildare, comprising in all about 238,500 acres, it consists of a series of contiguous morasses about 250 feet above the sea, and separated by ridges of dry ground, its East end (Clane Bog) being 17 miles West of Dublin. Average depth of peat 25 feet resting on clay and marl, it is traversed by the Grand canal, and in it the Barrow, Boyne and Brosna rivers have their sources.

ALLEN.—A Lough, a lake of province Connaught, County Leitrim, 9 miles North of Carrick, 7 miles in length, North to South, by 1 to 3 miles in breadth, 144 feet above the sea. It is generally regarded as the source of the Shannon. The town Drumshambo is on its South shore.

AMESTOWN.—A market village of County Waterford, 10½ miles Southwest Waterford, on a small bay. Population, 149.

ANTRIM.—The Northeastmost County of Ireland, Province Ulster, having North the Atlantic ocean, East the North Channel, dividing it from Scotland South and West. The counties of Down and Londonderry and Southwest Lough Neagh, separating it from Counties Tyrone and Armagh. Area, 1,164 square miles, or 745,177 square acres, of which 503,288 are arable, 176,335 uncultivated, 10,358 in plantations, 1,908 in towns and 53,288 in water, including part of Lough Neagh. Inhabited houses in 1851, 44,232. Population, 352,264. A third part of the surface, near the coast, is mountainous and rises in some places to 1,600 feet in height in the Southwest. Much of the soil is boggy. Chief rivers: The Bann, forming the West and the Lagan the South boundary. The famous Giant's causeway is on the North Coast of this County. Property is in large estates, but farms are small. In 1841 the total number of farms was 23,526, and of these 6,855 measured from 1 acre to 5 acres each, 4,220 from 15 to 30 acres each and 1,188 upwards of 30 acres. Most of the inhabitants, especially about Belfast, are engaged in spinning linen and cotton yarn, and in weaving. Salmon and other fisheries on the coast are important. The county is sub-divided into 14 baronies and **94** parishes. Carrickfergus is the Capitol, but the largest towns are Belfast, Lisburn and Ballymena. This County returns **two members to the House of Commons.**

ANTRIM.—An inland town and parish in County Antrim, on six mile water, near its mouth on Lough Neagh, on railroad, and 14 miles Northwest Belfast. Area of parish, 8,884 acres. Population, 5,182; of the town, including Masserene, 2,645. It has two good streets, a church, several dissenting chapels, a Union workhouse, a Courthouse in which general and petty sessions are held, and numerous schools. Manufactures of linen, calico, hosiery, paper bleaching and malting are carried on here. Meal and malt are sent to Belfast by Lough Neagh and the Lagan, and by rail. Markets on Thursday, and fairs January 1st., May 12th., August 1st., and November 12th. Near it are Antrim Castle, the seat of Lord Masserene, Shane's Castle, the residence of Lord O'Neil and the O'Neil family, and one of the most perfect of the round towers of Ireland.

ARDAGH.—A village and parish of County Longford, and 5½ miles Southeast of town of Longford. Area of parish, 11,417 acres. Population, 4,524; of village, 165. It has an old church, and was, until 1685, the seat of a bishopric; now united to Tuam. There are four other parishes of the same name, *as follows:* 4½ miles West Youghal, County Cork; 5 miles West Raith Keale, County Limerick; 2½ miles West Southwest Ballin, County Mayo; 4½ miles Northeast Nobber, County Meath.

AROARA.—A town of Ulster County, and 15 miles Northwest Donegal, head of Lochrus Bay. Population, 603.

ARDBRACCAN.—A parish of County Meath, 2½ miles West Northwest Navan. Area, 6,491 square miles. Population, 4,596. Ardb House, the seat of the Bishop of Meath, is one of the finest Episcopal residences in Ireland.

ARDEE, (ATHERDEE).—A town on the Dee, a municipal borough, town and parish of County Louth, Capitol Barony, on the Dee, 12 miles Northwest Drogheda. Area of parish, 4,885 acres. Population of parish, 6,392; of town, 3,679. It consists mostly of wretched cabins, but has some good houses, two old castles, (one now a Courthouse), a church of the 13th century, a spacious Roman Catholic Chapel, Union workhouse, dispensary, several schools and at one end of the town is a remarkable mound called

the Castle Guard. Corporation Revenue about £135 a year. General sessions in January and June. Petty sessions weekly. It has trade in malt and corn. Market Tuesday. Fairs, mostly for live stock, seven times a year.

ARDFERT.—A village and parish of Munster, County Kerry, near Ballyheigue Bay, four miles Northwest Tralee. Area, 6,797 acres. Population, 4,074; of village, 655. It was formerly a Bishop's see and an ancient parish; its Cathedral is now the Parish Church.

ARDFINNAN.—A village and parish of Munster, County Tipperary, on the Suir, 6½ miles Southwest Clonmel. Area of parish, 1,813 acres. Population, 1,214. The ruins of a castle built by Prince John in 1184 are still to be seen.

ARDGLASS.—A seaport town and parish; County Down, Ulster, situated on the Irish Sea, 6 miles Southeast Downpatrick. Area of parish, 1,137 acres. Population 1,433; of town, 1,166. It stands on elevated ground between two hills, and has many new and handsome residences frequented by visitors during the bathing season. A castellated mansion of the Chief proprietor is erected on a range of what were formerly spacious warehouses. This town having enjoyed a flourishing commerce during the Lancastrian dynasty. Trade now is principally in the herring and other fisheries, and the export of corn. It belongs to the port of Killough, one mile Southeast and has an inner cove for vessels of 100 tons, besides a large outer harbor for ships of 500 tons, protected by a pier with a lighthouse at its extremity. Market, Thursday, and fairs eight times annually.

ARDMORE.—A seaport town and parish of Munster, County Waterford, on Ardmore Head, 4 miles Northeast Youghal. Area of parish, 24,215 acres. Population 8,737; of town, 716. The principal business of the place is fishing. In its churchyard is a well preserved round tower, and its ruined church, a dormitory, a well, and a greatly venerated stone, all bear the name of St. Declan, reputed in early Christian times to have founded a monastery here.

ARDNAGEEHY.—A parish of Munster, County Cork, 5½ miles Southwest Rathcormack. Area, 16,335 acres. Population, 4,798.

ARDNAGLASS BAY.—An inlet West Coast of Ireland, Connaught, County Sligo. It extends inland for 6 miles, with an average breadth of 2 miles. It receives the Owenbeg River, and its head is the town of Ballysadore.

ARDNAREE.—County Mayo is that part of the town of Ballina, East of the River Moy.

ARDNURCHER; or HORSELEAP.—Is a parish of Westmeath and Kings Counties, Leinster, 4 miles West Northwest of Kilbeggan. Area, 12,012 acres. Population, 3,687. There are curious remains of an old castle here.

ADRAHAN.—A parish of Connaught, County Galway, 7 miles North by West of Gort. Area, 17,984 acres. Population, 4,191.

ARDREA.—(or ARDREE.) Are two parishes of Leinster. One in County Queens, comprising part of the town of Mountmellick. Area, 7,726 acres. Population, 5,185; another in County Kildare, 1 mile South of Athy. Area, 323 acres. Population, 205.

ARDSTRAW.—A parish of Ulster, County Tyrone, comprising the town of Newtown Stewart, and villages of Ardstraw and Douglas Bridge. Area, 44,974 acres. Population, 17,384. Danish Forts and Antiques are very numerous in this parish.

ARDTREA; or ARTREA.—A parish of Ulster, partly in County Tyrone; chiefly in County Londonderry and comprising part of the town of Moneymore. Area, 41,895 acres; of which 2,526 are water. Population, 25,546.

ARIGAL.—Is a mountain in Ulster, County Donegal. It is situated 7½ miles East of Guidore Bay. It is 2,462 feet in Height.

ARIGNA.—Is a district of Connaught, County Roscommon. It is on the West side of Lough Allen, 9 miles North of Carrick. Coal and Iron are found but not in paying quantities.

ARKLOW.—Is a Barony, situated in the Southeast of County Wicklow, Leinster; it includes nine parishes. The title is given to the House of Ormonde. Population, 25,263. Area 67,357 acres. ARKLOW, a parish and seaport town of Arklow barony, and the largest town in the County of Wickford. It is situated on the Ovoca River, near its mouth, in the Irish Sea. It is 13 miles South by East of Wicklow. Population of parish, 6,237; of town, 3,254. Area of parish, 8,127 acres. The town is divided into two parts; an upper and lower town. The upper town has a very good main street, but the lower town is composed chiefly of fisher's huts. The principal buildings are the Church, Roman Catholic and Methodist Chapels, Fever Hospital, Barrack on the site of an ancient castle and a bridge of nineteen arches over the Ovoca. It has numerous schools and a large number of boats are employed in herring and oyster fishing, and although its harbor is shallow and impeded by a bar, it has some trade in exporting Corn, Copper Ore and Fish; and in importing Coals and Provisions. There is a floating light situated on the South end of Arklow bank. Latitude, 52° 42' North. Longitude, 5° 57' West.

ARMAGH.—Is an inland county, Ulster; having on the North, Lough Neagh, East, the county Down, West, the Counties Tyrone and Monaghan and South, by Louth. Area, 328,076 acres; of which 265,343 are arable and the rest is divided into plantations, towns, &c.. Population 196,085; occupying 35,197 houses. The country is flat, except in the Southwest where Slieve Gallion rises to the height of 1,893 feet. The soil is fertile and well watered by the rivers; Callan, Blackwater, Ban and Newry-water. Some large estates belong to the Church, Nobility, &c., but, as a general thing the farms and properties are small. Average rent of land, 17 shillings per acre. Weaving is often combined with agriculture— Linen being the staple manufacture. The county is divided into 8 baronies and 28 parishes and parts of parishes. It returns two members to the House of Commons.

ARMAGH.—(The lofty field), Is a city and Parliamentary and municipal borough and parish near the River Callan and the Ulster Canal, 70 miles North by West from Dublin. Population, 12,654. Area, 4,607 acres. It is connected by the Ulster Railroad with Belfast. It is well built, chiefly of red marble. The streets diverge from the Cathedral down the sides of the hill and are clean, well lighted with gas and supplied with water. It has a Protestant Chapel of Ease, two Roman Catholic, two Methodist, one Independent and three Presbyterian Chapels, a County Court House, Prison, Infirmary, Lunatic Asylum, Grammer and various other Schools, five Banks, Tontine, News and Assembly Rooms, a Public Library with 14,000 volumes, Observatory with fine apparatus, Barrack for 800 men, Union Workhouse and a Public Promenade, called the mall. The Archbishop's palace near the city is plain but elegant. A few years ago about £12,000 worth of brown Linens were sold weekly on an average at its linen hall, and the average weekly sales of yarn, was £3,500. Armagh returns one member to the House of Commons. Assizes and quarter sessions are held here, besides a Manorial Court by the Archbishop for pleas of £10 and under. Tuesday is general Market day; Wednesday and Saturday for grain. There are twelve Fairs held here Annually. The Diocese comprises 118 parishes, chiefly in the Counties of Armagh and

Louth. Episcopal Revenue, (1833), £14,494. The Linen manufactories, to a large extent, are worked by steam, and in 1863, 1864 and 1865 the demand for their productions was very large, but since that time it has gradually fallen off.

ARRAN ISLES.—Is a sea-girt barony of County Galway, Connaught, and consists of a group of small islands. Inishmore, the largest and most Northerly, is situated in Latitude, 53° 7′ 38″ North. Longitude, 9° 42′ 22″ West. There is a Lighthouse on this island. Area, 11,287 acres. Population, 3,000. The soil of this group is very fertile, but dry in summer. The chief products are, Potatoes, Rye, Oats, fresh and cured Fish and Puffin's Feathers. Inishmore has many antiquities, and on its East coast is the village of Killeany. The islands give the title of Earl to the head of the Gore family. North Arran or Arranmore Island is populated by 1,000 souls, employed in agriculture and fishing. Area 4,335 acres. The Lighthouse on this island exhibits a fixed light.

ARTHURSTOWN.—Is a hamlet and seaport town of Wexford County, Leinster, situated on the East shore of Waterford Harbor, 7 miles East by South of Waterford. Population, 285. It has a small dock attainable by vessels of 100 tons, and it is an out-port to Waterford; having a little trade in the export of Fish and the import of Coal and Culm.

ASKEATON.—Is a town and parish of Limerick County, Munster, situated on the River Deel, two miles from its confluence with the Shannon and 17 miles West Southwest of Limerick. Area, 6,521 acres. Population, 4,438; of town, 1,862. Its Parish Church was that of a Commandery of Knights Templars founded in 1298. It has a Royal Chapel, several Schools, the ruins of a Castle and of a Franciscan Monastery. The river is navigable up to the town for vessels of 60 tons.

ATHBOY.—(The yellow ford), Is a town and parish of County Meath, Leinster, situated on the Athboy River, 6 miles Northwest Trim. Population, 5,365; of town 1,826. Area, 11,884 acres. It consists of a principal street, with a Church, Chapel, Session House, a large School House and a Widow's Alms House supported by the Earl of Darnly.

ATHENRY.—Is a town and parish of County Galway, Connaught, it is situated 13 miles East of Galway. Area, 24,952. Population, 5,988; of town 1,236. It is poor and dull, but is one of the oldest towns in the county, and parts of its ancient walls, gates, &c., are still traceable.

ATHLEAGUE.—Is a parish of Roscommon and Galway Counties, Connaught. It is situated on the River Suck, which is here crossed by a series of bridges. It contains a small village of 631 inhabitants. Area, 13,012 acres. Population, 5,087.

ATHLONE.—(ATH-LUAN, ford of the moon), Is a fortified town and parliamentary borough and parish of County Westmeath, Leinster, and County Roscommon, Connaught. It is situated 1½ miles South of Lough Ree and 70 miles West of Dublin. Latitude, 53° 25′, 24″ North. Longitude, 7° 56,′ 29″ West. It contains 2 parishes. St. Mary's or Athlone has an area of 11,456 acres. Population of the town 6,207. St. Peters on the right branch of the Shannon has an area of 7,617 acres. Population, 3,460. During the war with France it was defended Westward by works covering 15 acres and had barracks for 1,500 men in its old and strong castle. The town is ill-built and inconvenient, and contains 2 parish Churches, various Chapels, a Court House, Bridewell, Union Workhouse and many Public Schools; one endowed with 470 acres of land. In and near the town are various Distilleries, Breweries, Tanneries, Soap Works and Flour Mills. A brisk trade is carried on with Shannon Harbor and Limerick by Steamers, and with Dublin by Canal. This Town is the Military Headquarters for the West of Ireland and is mostly supported by the expenditure of the Garrison.

ATHY.—(ATHLEGAR, the Western ford,) is a Market Town of County Kildare, Leinster. It is situated on the Barrow River and on an arm of the Grand Canal, and on the Carlow Railroad 42½ miles Southwest of Dublin. It is the seat of the County Summer Assizes, and has trade in Corn, Butter and Malt with Dublin, New Ross and Waterford. It contains a Church, Parish and other Schools, Cavalry Barracks, a Court House, Union Work House, Fever Hospital, a Police Barrack in the remains of its ancient Castle, and near the Town is the County Jail. The Town stands partly in the two parishes of St. John and St. Michael, which have a united area of 422 acres. Population 9,396. Markets, Tuesday and Saturday; Fairs, March 17, April 25, June 9, July 25, October 10 and December 11.

ATTYMASS.—(or ATTIMASS), Is a parish of County Mayo, Connaught. It is situated 3½ miles North Foxford. Area, 11,154 acres. Population, 3,435. The soil is barren and marshy.

AUBURN.—Is a village of County Westmeath, Leinster. It is supposed to have been the village of which Goldsmith wrote in his "Deserted Village," it was originally called Lishoy. It is situated near Lough Ree, 6 miles North of Athlone.

AUGHALOO.—(or Aughloe), Is a Parish of County Tyrone, Ulster. It contains the town of Caledon. Caledon Hill, the seat of the Earl of Caledon, is also here. Area, 19,583 acres. Population of Town, 9,867; of Parish, 8,821.

AUGHAVAL.—(or Oughaval), Is a Parish of County Mayo, Connaught, and consists of the towns of Westport and Westport-Quay. Area, 33,695 acres. Population, 13,441. The country is mostly mountainous and boggy. Population of Rural districts, 8,529.

AUGHAVEA. Is a parish of County Fermanagh, Ulster. It is situated 12 miles Northwest of Clones. Area, 17,142 acres. Population, 6,730.

AUGHER.—Is a town of County Tyrone, Ulster. It is situated 2 miles Northeast of Clogher. There is a castle of the same name here. Population, 753.

AUGHMACART.—Is a parish of Queens County, Leinster. It is situated 4½ miles Southwest of Castle Durrow. Area, 9,601 acres. Population, 3,667.

AUGHNACLOY.—Is a small town of Tyrone County, Ulster, situated 11 miles North of Monaghan. It is in the parish of Carrateel. Population, 1,841. They have a market here every Wednesday.

AUGHNAMULLEN.—Is a parish of County Monaghan, Ulster, and is situated 2½ miles South of Ballybay. Area, 30,710 acres, which includes numerous Loughs. Population, 18,219.

AUGHNISH.—(or Aghnish), Is a parish of Donegal County, Ulster, and comprises part of the town of Ramelton. Population, 4,974; mostly engaged in manufacturing Linen. Area, 9,195 acres.

AUGHNISH.—Is a Village of Galway County, Connaught, and is situated on the South side of Galway Bay. Population, 312.

BADONY.—(or Bodony), There are two parishes by this name, one, called the Lower, is situated in Tyrone County, Ulster, 4½ miles East of Newtown-Stewart. Area, 47,920 acres. Population, 7,784. The other called the Upper, is situated 10 miles Northeast of Newtown-Stewart. Area, 38,208 acres. Population, 5,822. The Lower contains the village of Gortin.

BAGENBUN HEAD.—Is a cape at the entrance to Bannon Bay, in the County Wexford, Leinster. It was here that Earl Strongbow made his descent on Ireland in 1170.

BAGNALSTOWN.—Is a Town of Carlow County, Leinster, and is situated 10 miles South of Carlow Town. It is on the river Barrow. Population, 2,225. Adjoining it are the mansions of Dunleckney and Bagnalstown.

BAILIEBOROUGH.—Is a town and parish of Cavan County, Ulster, and is situated 17 miles Southeast of Cavan at the head of the Blackwater River. The Castle of the same name is on the site of the ancient Castle of Touregie. Area of Parish, 12,416 acres. Population of Parish, 6,984; of Town, 1,203.

BALBRIGGAN.—Is a Maritime and Chapelry of Dublin County, Leinster, parish of Balrothery. It is a thriving town, and is situated on the Irish Sea, 18½ miles North Northeast of Dublin. Population, 2,959. It is a favorite watering place, and contains many handsome buildings. The harbor is protected by a Quay, on which there is a Lighthouse with a fixed light. Latitude 56°, 36',45", North. Longitude, 6°, 11', West.

BALLA.—Is a village of Mayo County, Connaught; situated 8 miles Southeast of Castlebar. It contains about 600 inhabitants.

BALLAGHADERIN.—Is a town of County Mayo, Connaught; situated near the River Lung, and 30 miles East Northeast of Castlebar. Population, 1,342. There is a small Infantry Barrack here.

BALLAGHMORE.—Is a village of Queens County, Leinster. It is near Roscrea, and contains the ruined Abbey of Monaincha

BALLINA —Is a town of County Mayo. Connaught. It is situated on the left bank of the river Moy, 18 miles North Northeast of Castlebar. This is the third largest town of this county, and contains a large number of handsome public buildings and manufactories. Population, 5,186, exclusive of Ardnaree. Its Salmon fishery ranks next in importance to that of the Bann. General Sessions in July; Petty Sessions on Tuesday. Market day, Monday. Fairs, May 12 and August 12.

BALLINABOY.—Is a Parish of County Cork, Munster, between the City of Cork and Kinsale, and comprises the villages of Ballinahassig and Ballytrooleen. Area, 7,973 acres. Population, 2,749.

BALLINACARRIG.—Is a parish of County Carlow, Leinster,

and situated 1¼ miles South Southeast of Carlow. Area, 2,605 acres. Population, 692.

BALLINACARRIG.—Is a hamlet of County Cork, Munster, situated on the river, and 8 miles West of Baudon, with ruins of a Castle of the 16th century.

BALLINACOURTY.—There are two parishes of this name ; the first is in County Galway, Connaught, at the head of Galway Bay, and the second is in County Kerry, Munster, on the North side of Dingle Bay, 10 miles Northeast of Dingle. Area, of first, 6,293 acres. Population, 3,407. Area, of second, 5,318 acres. Population, 1,472.

BALLINAHAGLISH.—There are two parishes by this name; one in County Mayo, Connaught, 2½ miles South Southeast of Ballina on the River Moy. Area, 12,659 acres. Population, 5,397; another, in County Kerry, Munster, situated 5 miles West of Tralee. Area, 3,006 acres. Population, 2,147; comprises the villages of Chapeltown and Kilfinura.

BALLINAHINCH.—Is the name of a barony, parish, domain, lake, river, seat and ruined castle of the district of Connemara, Connaught. The seat of the proprietor of the district is 37 miles North Northwest of Galway. The barony with an area of 191,433 acres, comprises the mountain group of the "Twelve Pins" and the seaport of Clifden. Population, 33,465. Second, is a small town of County Down, Ulster, 10 miles East of Dromore. Population, 911. It comprises most of the Chapels and Schools of the parish Magheradroll. In June 1798 a battle was fought here between the Irish Insurgents and Royal Troops.

BALLINAKILL.—Is a town of Queens County, Leinster, 11 miles South of Maryboro. Population, 1,540. It has a handsome Church and Roman Catholic Chapel, and remains of an old Castle.

BALLINAMORE.—Is a village of County Leitrim, Connaught, 13 miles Northeast of Carrick. Population, 946. It has a Church, Roman Catholic and Methodist Chapels, a Session House and a Bridewell.

BALLINAMUCK.—Is a village of County Longford, Leinster. It is situated 11 miles North Northeast of Longford. Here the French troops under Gen. Humbert, surrendered to the English troops Sept. 8th, 1798.

BALLINASCREEN.—Is a parish of Londonderry County,

Ulster, and is 8 miles West Southwest of Tobermore. Area, 32,492 acres. Population, 8,384.

BALLINASKELLIGS.—(or BALLINSKELLIGS) Is a bay at the entrance between Hog Head on East, and Bolus Head on North, County Kerry, Munster. Breadth, 5 miles.

BALLINASLOE.—Is a town of Counties, Galway and Roscommon, Connaught. It is situated on the river Suck, 22 miles South of Roscommon. Population, 4,934. The Suck River divides it into two parts, connected across an Island by a causeway and two bridges. The town is very neatly built and has a curious church, and several fine buildings. The largest fair in Ireland for the sale of cattle and sheep is held here annually. About 50,000 to 60,000 cattle, and from 6,000 to 7,000 sheep, are the average numbers sold at each fair. This is also a station of the Galway Militia.

BALLINCALLA.—(or BALLINCHOLA) Is a parish of Counties, Galway and Mayo, Connaught, 4 miles West Southwest of Ballinrobe. Area, 15,195 acres. Population, 2,165.

BALLINCOLLIG.—Is a town of County Cork, Munster; situated 5 miles West of Cork, on the River Lee. Population, 1,287. There is a large Military Barrack here.

BALLINCUSLANE.—(or BALLYCUSLANE) Is a parish of County Kerry, Munster; situated 10 miles Northwest of Mill Street. Area 39,740 acres; mostly mountainous. Population, 5,701.

BALLINDERRY.—There are two parishes by this name; one in County Antrim, Ulster, 4½ miles North of Moira. Area, 10,891 acres. Population, 5,679. Another, in County Tyrone, Ulster, 4 miles Southeast of Moneymore. Area, 8,178 acres. Population, 1,189. The village by this name is in Wicklow County, Leinster, 2 miles Northwest Rathdrum. There is also a Hamlet and River by the same name.

BALLINDOON.—Is a parish of Galway County, Connaught, and 45 miles Northwest of Galway. Area, 20,033 acres. Population, 5,615.

BALLINGADDY.—Is a parish of County Limerick, Munster, 2½ miles East by South from Kilmallock. Population, 1,761.

BALLINGARRY.—Is a town and parish of Limerick County, Munster, situated 17 miles Southwest of Limerick. Area of parish, 17,737 acres. Population of parish, 8,679; of town, 1,690. There are ancient ruins here. There is also a village and parish of the same name in County Tipperary, Munster, situated 20 miles Northeast of Clonmel. Area of parish, 13,714 acres. Population of parish, 7,062 ; of town, 643 ; employed in the Slievedargy Coal Mine. Two other parishes of this name are in Counties, Limerick and Tipperary.

BALLINLANDERS and BALLINLOE.—Are two parishes of Ireland. The first is in County Cork, Munster, 5 miles West by South from Tallow ; the second is in County Limerick, Munster, 10 miles East Southeast from Kilmallock.

BALLINROBE.— Is a town and parish of County Mayo, Connaught ; situated on the river Robe, 3 miles from its entrance into Lake Mask, and 16 miles South Southeast of Castlebar. Area of parish, 26,903 acres. Population of parish, 11,150 ; of town, 2,678. It is well built and contains many handsome buildings. General Sessions in June and December; Petty Sessions and Market, Monday. Fairs, Whit Tuesday and December 5.

BALLINTEMPLE.—Is a parish of County Cavan, Ulster, and situated 6 miles South Southwest of Cavan. Area, 10,658. Population, 5,341.

BALLINTOBBER.—Is a village and parish of County Roscommon, Connaught, and 12 miles Northwest of Roscommon. It has the ruins of a fine Castle. Area, 6,352 acres. Population, 2,616. This is also the name of two baronies in the same County, and of a parish in County Mayo.

BALLINTOGHER.—Is a township of County Sligo, Connaught, and is 7 miles Southeast of Sligo. Population, 234.

BALLINTOY.—Is a maritime village and parish of County Antrim, Ulster, and is 4 miles North of Ballycastle. Area, 12,654 acres. Population, 4,816.

BALLYBAY.—Is a town and parish of County Monaghan, Ulster, and is 8 miles South Southeast of Monaghan, on the road to Dublin. Area, 3,641 acres. Population of parish, 6,606 ; of town, 1,768. It is a thriving town, and has a Public Library of about 1,000 volumes.

BALLYBOFEY.—Is a town of County Donegal, Ulster, parish of Stranorlar. It is situated on the river Finn, 14 miles West Southwest of Lifford. Population, 784. It has a Union Workhouse and is the centre for the retail trade of the district about here.

BALLANBOY.—Is a parish of County Kings, Leinster, and is 10 miles Northeast of Birr. Area, 14,274 acres. Population, 4,758. Market on Saturday. There are seven Annual Fairs held here.

BALLYBUNNION.—Is a township of County Kerry, Munster, on the river Shannon, 17 miles North of Tralee. Population, 271. It is much resorted to for Sea-bathing, and near it are many Maritime Caves, one of which is from 70 to 80 feet in height.

BAYLLYBURLEY and BALLYCALLEN.—Are two parishes of Ireland. The former is in County Kings, Leinster, on the Grand Canal, 12 miles East Northeast of Philipstown. The latter is in County Kilkenny, Leinster, and 4½ miles West Southwest of Kilkenny.

BALLYCASTLE.—Is a seaport town of County Antrim, Ulster; situated in a bay opposite Rathlin Island, and 5 miles Southwest of Fairhead. Population, 1,697. It is a handsome and well built town and has many fine buildings, but is very quiet. There was £150,000 spent for the improvement of its harbor, which is now filled with sand. It has a small Linen manufactory, a Salmon fishery, and traffic in Rathlin Ponies. There is a maritime village of same name in County Mayo, Connaught, 31 miles North of Castlebar. Population, 798. It is a Coast Guard station and is resorted to for Sea-bathing.

BALLYCLARE.—Is a Market town of County Antrim, Ulster, 11 miles North of Belfast. Population, 847.

BALLYCONNELL.—Is a town of County Sligo, Connaught, 9 miles Northwest of Sligo. Population, 553. There is a town of same name in County Cavan, Ulster, 13 miles Northwest of Cavan. Population, 387.

BALLYCOTTON.—Is a bay and village of County Cork, Munster, and is situated 20 miles Southeast of Cork. Population, 449.

BALLYEASTON.—Is a parish of County Antrim, Ulster, and its village is 1½ miles North of Ballyclare. Area of parish, 13,799 acres. Population, 265.

BALLYFIN.—Is a chapelry and a seat of the Pole family 4 miles North of Maryborough.

BALLYHAISE.—Is a town of County Cavan, Ulster, and is 4 miles North Northeast of Cavan. Population, 704.

BALLYHEIGH.—(or BALLYHEIGUE). Is a parish and township of County Kerry, Munster, on Ballyheigh Bay, which is an inlet of

Tralee Bay, and is 9 miles Northeast of Tralee. Area of parish, 11,261 acres. Population, 4,795.

BALLYJAMESDUFF.—Is a town of County Cavan, Ulster, and is 11 miles Southeast of Cavan. Population, 1,071.

BALLYKEAU. Is a parish of County Kings, Leinster, and is 3½ miles North Northwest of Porterlington. Area, 12,201 acres. Population, 2,445.

BALLYLONGFORD.—Is a town of County Kerry, Munster, and is 5 miles West Southwest of Tarbet, on the estuary of the Shannon river. In the vicinity are the ruins of Lislaghtin Abbey.

BALLYLOUGHLOE.—Is a parish of County Westmeath, Leinster, and is 6 miles East of Athlone. Area, 13,577 acres. Population, 4,793.

BALLYMACELLIGOTT and BALLYMYAECK.—Are two parishes. The first is in County Kerry, Munster, 5 miles South Southeast of Tralee. Area, 14,018 acres. Population, 4,058. The second is in County Tipperary, Munster, and is 4½ miles East of Nenagh. Area, 9,713 acres. Population, 3,178.

BALLYMAHON.—Is a town of County Longford, Leinster, parish of Shruel, and is on the river Inny, 11 miles South by East of Longford. Population, 1,220. It consists of one principal street, and contains some fine buildings and a handsome Bridge of five arches. Quarter and Petty Sessions, and Weekly Market for Corn. Cattle Fair, May 11, and other Fairs Ash Wednesday, August 11 and November 21.

BALLYMOSCANLON.—Is a parish of County Louth, Leinster, and is 3 miles Northeast of Dundalk. Area, 15,997 acres. Population, 6,674. This is a very handsome county.

BALLYMENA.—Is a town of County Antrim, Ulster, on the river Braid, 21 miles Northwest of Carrickfergus. Population, 6,009. This is a well built and handsome town, doing a brisk business, and has a good Market and other buildings. It exports Linen and Potatoes, and has numerous bleaching grounds and Mills of various kinds. General Sessions every June and January, alternately with Ballymoney. Petty Sessions on alternate Tuesdays. Market, on Saturday for Linens ; and two other Markets weekly for Grain and Provisions. Fairs, July 26 and October 21.

BALLYMONEY.—Is a town and parish of County Antrim, Ulster, and is 17 miles Northwest of Ballymena. Area of parish, 22,676 acres. Population of parish, 11,727 ; of town, 2,490. It is irregularly built on an eminence, and has a Town Hall, School House and various places of Worship and Public Buildings. Its principal trade is in Matting, Tallow, &c. Market on Thursday. Fairs, May 6, July 10 and October 10. There is a parish and township of the same name in County Cork, Munster, which is 23 miles Southwest of Cork. Population of parish, 3,733.

BALLYMORE.—There are several parishes and villages by this name. The first, in County Westmeath, Leinster, 13½ miles West Southwest of Mullingar. Population, 3,487. The second, in County Wexford, Leinster, and 15 miles North Northeast of Wexford. Population, 568. The third, in County Armagh, Ulster, and 20 miles South Southwest of Armagh. Population, 11,520. The fourth, (EUSTACE,) is in County Kildare, Leinster, and is on the river Liffey, 19 miles Southwest of Dublin. Population, 2,219; of village, 936.

BALLYMOTE.—Is a village of County Sligo, Ulster. Population, 839. It has a large Roman Catholic Church and ruins of a Castle built in 1300, and of a Franciscan Monastery.

BALLYGOVEY.—Is a parish of County Mayo, Connaught, and is 6¾ miles North Northwest of Ballinrobe. Area, 27,622 acres. Population, 4,605.

BALLYRAGGET.—Is a town of County Kilkenny, Leinster, parish of Donaghmore. It is situated on the river Nore, 10 miles North Northwest of Kilkenny. Population, 1,577.

BALLYSADARE.—Is a thriving town and parish of County Sligo, Connaught, and is finely situated at the head of Ballysadare Bay, a branch of Sligo Bay. Area of parish, 16,025 acres. Population of parish, 7,822; of town, 869. The river here rolls over shelving rocks, turning several Mills, and at the foot of its lowest fall is a harbor with safe anchorage.

BALLYSAX and BALLYSCULLION.—Are two parishes. The first is in County Kildare, Leinster, and 3 miles Southwest of Kiscullen Bridge. The second is in County Antrim, Ulster, and 4 miles Northeast of Magherafelt. Area, 12,750 acres. Population, 6,979.

BALLYSHANNON.—Is a sea port town of County Donegal, Ulster, of which it is the principal town. It is situated on the river Erne at its mouth in Ballyshannon Bay, 25 miles Northwest of Enniskillen. Population, 3,513. It consists of three steep and irregular streets on one side of the river, across which it communicates with the wretched suburb of Purt. There are several public buildings here and the ruins of the ancient Castle of the Earls of Tyrconnell. The harbor is very poor. The only Newspaper in the County is published here.

BALLYTORE.—Is a town of County Kildare, Leinster; is 11 miles South Southeast of Kildare. Population 441. Here Edward Burke received the rudiments of his education.

BALLYVOURNEY.—Is a parish of County Cork, Munster; its village is situated 7½ miles West of Macroom. Area, 26,603 acres. Population, 4,466.

BALLYWALTER and BALLYWILLIN.—Are two parishes. The first is in County Down, Ulster, with a maritime village 7 miles Southeast of Donaghadee. It has valuable Slate quarries. The second is in County Londenderry, Ulster, 3½ miles North Northeast of Coleraine.

BALTIMORE.—Is a seaport town of County Cork, Munster; situated on a small bay of the Atlantic, 47 miles Southwest Cork, Latitude, 51°, 29', North. Longitude, 9°, 20', West. Population, 168. It has a large coast trade.

BALTINGLASS.—Is a town and parish of Counties, Wicklow and Carlow, Leinster. It is situated on the river Slaney, 12 miles Northeast of Carlow. Area of parish, 5,273 acres. Population of parish, 4,436; of town, 1,928. The town is very meanly built, has a Bridewell and Infirmary with some bleach fields, remains of an Abbey of the Twelfth century, and a castle of the Earl of Aldborough, chief proprietor.

BANAGHER.—Is a town of County Kings, Leinster, parish of Reynagh. It is situated on the Shannon river which is here crossed by a bridge 400 feet long and guarded by batteries, and is 21 miles West Southwest of Tullamore. Population, 2,827. It consists of one long street with some fine buildings. Market on Friday for Corn. Fairs, September 15 and three successive days, October 28 and November 8. There is a parish of same name in County Londonderry, Ulster; 16 miles Southeast of Londonderry. Area, 32,475 acres. Population, 5,810.

BANBRIDGE.—Is a town of County Down, Ulster; parish of Seapatrick, on the river Bann, 7 miles Southwest of Dromore. Population, 3,324. The town is neat and thriving, and has a handsome new Church, several Dissenting Chapels, two Market Houses, and a Brown Linen Hall. This is the principle seat of the

Linen trade of the country and has extensive Cloth and Thread factories, bleaching grounds, and Chemical Works. Market on Monday. Fairs, 17 times annually; that on November 16th being a large Horse Fair.

BANDON.—(or BANDONBRIDGE). Is a town of County Cork Leinster, on the Bandon river, 15 miles Southwest of Cork. Population, 6,927. It occupies declevities on both sides of the river, and it has a number of public buildings, Bandon returns one member to the House of Commons. Market on Saturday, Fairs, 15 times annually. Both markets and fairs are toll free.

BANDON.—Is a river in County Cork, Leinster. It rises in the Cranberry Mountains, near Dunmanway, flows East to Innishannon and thence Southeast to the Atlantic Ocean, forming the harbor of Kinsale. Length, 40 miles; 15 miles of which it is navigable for vessels of 200 tons.

BANGOR.—Is a seaport town and parish of County Down, Ulster; is situated on Belfast Lough, 12 miles East Northeast of Belfast. Area, 17,027 acres. Population of parish, 10,060; of town, 3,116. The town is frequented as a bathing place. It contains various places of Worship, Savings Bank, Factories, &c. Market on Tuesdays Fairs, January 12, May 1, August 1 and November 22. Here was once a famous Monastery said to have been destroyed by the Danes in 820, and near the town is Bangor Castle, the seat of the Earl of Bangor, chief proprietor.

BANNON.—Is a parish of County Wexford, Leinster; is 18 miles Southwest of Wexford on Ballyteigue Bay. Here was formerly a town of same name, but since the Seventeenth Century it has become totally covered with sand.

BANN.—There are two rivers by this name in the North of Ireland, one flowing into and the other flowing out of Lough Neagh.

The UPPER BANN rises in the Mourne Mountains, flows through the Counties, Down and Armagh in a Northwesterly direction and joins Lough Neagh on the south side. It communicates with the Newry Canal. Banbridge, Gilford and Portadown are on its banks, and at the latter it becomes navigable for vessels of 60 tons.

The LOWER BANN issues from the North side of Lough Beg, flows North Northwest between the Counties Londonderry and Antrim and enters the Ocean 4 miles Southwest of Portrush after a course of 40 miles. Portglenone, Kilroa and Coleraine are on its banks, and though impeded by sand banks it is navigable for vessels of 200 tons from the sea to near the latter town. Its Salmon and Eel fisheries are important. Bann is also the name of an affluent of the river Slaney, County Wexford, Leinster.

BANTRY.—Is a seaport town of County Cork, Munster, parish of Kilmocoge, near the head of Bantry Bay, 44 miles West Southwest of Cork. Latitude, 51°, 41' North. Longitude, 9°, 27' West. Population, 4,082. It has some trade. It is near Seacourt, seat of the Earl of Bantry.

BANTRY BAY.—Is a large bay in the South of Ireland, County Cork, Munster, and is one of the finest harbors in Europe. It extends for 25 miles inland. There are some islands here, among which are Bear and Whiddy's Islands. There is from 20 to 40 feet of water and the anchorage is good everywhere.

BARRAGH.—Is a parish of County Carlow, Leinster, 3 miles Northwest of Clonegall. Area, 12,296 acres. Population, 3,742. Mount Leinster is 2,610 feet in elevation.

BARROW (or BONRAGH.)—Is a river of Ireland, second in importance. It rises in the Slievebloom Mountains, Leinster, flows generally Southward, and after a course of about 90 miles joins the Suir. It divides the Counties, Kildare, Carlow and Wexford on the East from Counties, Kilkenny and Queens on the West. The towns Portarlington, Monasterevan, Athy, Carlow, Craig and New Ross are on its banks. It is navigable for large ships from the sea to Ross, and for barges to Athy 60 miles above its mouth, where it is joined by a branch of the Grand Canal.

BEAR (or BERE) ISLAND.—Is a rocky Island of the West coast of Ireland, County Cork, Munster, in Bantry Bay, 13 miles West of Bantry. Length, 6 miles. Breadth, (average,) 1½ miles. On it is the township of Ballinacallagh, and it shelters Bearhaven on the North side of the bay.

BECAN (or BEKAN).—Is a parish of County Mayo, Connaught; is situated in the South of the barony of Costello. Area, 20,303 acres. Population, 5,589.

BECTIVE.—Is a parish of County Meath, Leinster, and is 4 miles Northeast of Trim. Area, 3,389 acres. Population, 602. The ruins of Bective Abbey, founded A. D. 1146, are here.

BELFAST.—Two baronies of County Antrim, Ulster. (Upper and Lower). The upper barony extends to the Southeast extremity of the County. Area, 35,898 acres. Area of lower barony, 56,993 acres.

BELFAST.—Is a parliamentary and municipal borough and seaport town of County Antrim, Ulster. It is situated at the head of Belfast Lough, where it receives the river Lagan, and is about 12 miles from the Irish Sea and 86 miles North Northeast of Dublin. Latitude 54° 36' 8'' North. Longitude 5° 55' 53'' West. Population, about 100,000. The town is well built, paved, drained and lighted, and is equaled by few other manufacturing towns in the empire. There are many fine public buildings, and the city contains no less than 25 Churches of various denominations. There are 10 Newspapers published in the borough. Belfast is the principal depot for the Linen trade, and is the chief seat of the Cotton manufactories of Ireland.

It has a large number of Linen and Cotton mills, mostly worked by steam power; extensive Distilleries, Breweries, Foundries, Shipyards, etc., and there are numerous large Bleaching grounds in the vicinity. The inland trade is greatly facilitated by a canal connecting the river Lagan with Lough Neagh, and by a railroad to Armagh 25 miles to the Southwest, and one to Antrim and Randalstown on the Northwest. There is also regular communication by steamers to London, Glasgow, Liverpool and Dublin. Vessels drawing 15 feet approach the docks, but those of heavier draft discharge at Garmoyle, 4 miles below. Friday is market day. Fairs, August 12th, and November 8th. General and petty sessions are held in the borough, which has a very fine police force. There are many handsome residences in the vicinity, among which is that of the Marquis of Donegal, who is lord of the manor.

It returns two members to the House of Commons. The borough was incorporated by James II. BELFAST-LOUGH is an inlet of the North Channel, 12 miles in lenght Northeast to Southwest, and 7½ miles in width at its entrance.

BALLAGHY.—There are two villages by this name. One in County Londonderry, Ulster; 6 miles North Northeast of Magherafelt. Population, 739. Another, in County Sligo, Connaught, 7 miles Northeast of Swineford. Population, 292.

BELMULLET.—Is a small seaport town of County Mayo, Connaught, situated on Blacksod Bay, 11 miles West Northwest of Bangor. Population, 637. It is a thriving town and has a coast guard station.

BELTURBET.—Is a municipal borough and market town of County Cavan, Ulster, and is 8½ miles North Northwest of Cavan on the river Erne. Population, 2,070. It has an Alms House and ruins of an ancient stronghold.

BENBURB.—Is a village of County Tyrone, Ulster, and is on the Blackwater river, 5 miles Northwest of Armagh. Population, 330. It has a ruined Castle. Benburb Falls are here and are very picturesque. The river runs close by the old Castle and makes power for four large Mills, Linen Manufactory and Foundry on the County Armagh side of the river.

BENNABEOLA (TWELVE PINS of,)—Are a group of mountains of County Galway, Connaught. Bencorr and Benbaun are the two loftiest, being 2,336 and 2,395 feet above the sea.

BETAGSTOWN.—Is a village of County Meath, Leinster; is situated on the coast 3 miles East Southeast of Drogheda on the Dublin and Drogheda Railway.

BILLY.—Is a parish of County Antrim, Ulster; is situated 8 miles Northeast of Coleraine. Area, 17,330 acres. Population, 7,277.

BINGHAMSTOWN (or SALEEN).—Is a maritime village of County Mayo, Connaught, on the West side of Blackwater Bay, 2½ miles South Southwest of Belmullet. Population, 436.

BIRR (or PARSONSTOWN).—Is a market town of County Dublin, Leinster; it is 62¼ miles West Southwest of Dublin. Population, 5,540. It is a well built town. A Castle belonging to the Earl of Rosse is here. It was known under the name of Biorra in the 6th Century, and in the 9th was a stronghold of the O'Carrols. In 1620, Sir W. Parsons, ancestor of Lord Rosse, received a grant of the town and adjoining estate from James I.

BLACKHEAD.—There are two capes of this name. One in County Antrim, Ulster, on the North entrance to Belfast Lough. Latitude, 54°, 46' North. Longitude, 5°, 42' West. Another in County Clare, Munster, on the South side of Galway Bay. Latitude, 53°, 9' North. Longitude, 9°, 16' West.

BLACK RIVER.—Is a tributary to the river Suir

BLACK ROCK.—Is a town of County Dublin, Leinster, 4 miles Southeast of Dublin, on the Dublin and Kingstown Railroad, and on the South shore of Dublin Bay. Population, 2,372. It is a place of resort for sea bathing. There are also several villages by this name. One in County Louth, Leinster, at the head of a small bay and 2½ miles Southeast of Dundalk. Population, 507; another in County Cork, Munster, 3 miles East of Cork, on a branch of the Lee. Population, 303. It has many handsome villas, among which is Castle Mahon, the residence of Lady Chatterton; a Nunnery, Blackrock Castle and many other antique edifices.

BLACKSOD BAY.—Is situated on the coast of County Mayo, Connaught. Latitude of entrance, 54°, 5' North. Longitude, 10° West.

BLACKSTAIRS.—Is a mountain range, and is part of the boundary between Counties, Carlow and Wexford. Mt. Leinster, 2,610 feet in height, is the highest peak.

BLACKWATER.—There are two rivers by this name. One rises in County Cork, Munster, about 16 miles Northeast of Killarney, flows Southward and enters the sea at Youghal. Length estimated to be 100 miles. The tide rises to Cappoquin, to where it is navigable. Mallow, Fermoy, Lismore and Youghal are on its banks. The other rises in Counties, Tyrone and Armagh, Ulster, and empties into Lough Neagh. Caledon and Charlemont are the principal towns on its banks.

BLACKWATERTOWN.—Is a village of County Armagh, Ulster, on the Blackwater river, 2½ miles South Southwest of Moy. Population, 369. There was a fort here famous in the time of O'Neill, in the 16th Century.

BLARNEY.—Is a village of County Cork, Munster, 4 miles Northwest of Cork. Population, 253. The scenery here is considered to be as fine as any in Ireland and has been rendered famous in song. The town itself is of small account, the greater part being in ruins. In its Castle, formerly the property of the Earls of Clancarty, is reported to be the famous "Blarney Stone," the kissing of which is said to impart that peculiar style of speech called "blarney." Fairs, Easter Monday and Tuesday, June 5 and 6, September 18 and November 11.

BLASKET ISLANDS.—Are a group of rocky islands on the West coast of Ireland, County Kerry, Munster, at the entrance to Dingle Bay.

BLENNEVILLE.—Is a small seaport town of County Kerry, Munster, on Tralee Bay, 1½ miles Southwest of Tralee. Population, 225.

BLESSINGTON.—Is a parish and market town of County Wicklow, Leinster, 18 miles Southwest of Dublin. Area, 15,780 acres. Population, 2,168; of town, 466. It is neatly built and has a commodious Church, Market House and Police Barrack. Weekly market on Fridays. Fairs, May 12, July 15 and November 12. It gave the title of Earl to the Gardner family; the widow of the last of whom was the well-known authoress; the late Countess of Blessington.

BLOODY FARELAND.—Is a promontory of County Donegal, Ulster, on its Northwest coast, 5 miles West Southwest of Innisboflin.

BOA ISLAND.—Is the largest island of Lough Erne, County Fermanagh, Ulster. Area, 1,400 acres.

BOHERMEEN.—Is a village of County Meath, Leinster, 4½ miles West Northwest of Navan, on the road to Kells. Population, 831.

BOHOE.—Is a parish of County Fermanagh, Ulster, 7½ miles West Northwest of Inneskillen.

BOHOLA.—(or BUCHOLLA). Is a parish of County Mayo, Connaught, 6½ miles Southeast of Foxford. Area, 8,674 acres. Population, 4,301.

BONMADNON.—(or BUNMAHON). Is a maritime village of County Waterford, Leinster, and 14 miles Southwest of Waterford at the mouth of the river Mahon. Population, 1,771. Near it are the Copper mines of Knockmahon.

BOOTESTOWN.—Is a parish of County Dublin, Leinster, 3½ miles Southeast of Dublin. Population, 3,318.

BORRIS.—(or BURRIS-IDRONE). Is a village of County Carlow, Leinster, 16 miles South of Carlow. Population, 950. In it is Borris Castle, the seat of Thomas Kavanagh, Esq.

BORRIS in OSSORY. Is a market town of County Queens, Leinster, 7 miles East Southeast of Roscrea. Population, 821. It was formerly a military position of some strength, and has a neat Court House.

BORRISLEAGH.—Is a parish of County Tipperary, Munster, 3½ miles Southeast of Thurles. Area, 10,940 acres. Population, 3,372.

BORRIS-O-KANE.—Is a town and parish of County Tipperary, Munster, 12 miles Southwest of Parsonstown. Area of parish 5,128 acres. Population, 3,175; of town 1,625.

BORRIS-O-LEAGH.—Is a small town of County Tipperary, Munster, 6 miles South Southwest of Templemore. Population, 1,438.

BOWRNEY.—(or BOURCHIN) Is a parish of County Tippperary, Munster, 4 miles South by West of Roscrea. Area, 12,981 acres. Population, 4,620.

BOVEVAGH.—Is a parish of County Londonderry, Ulster, 4½ miles North Northwest of Dungiven. Area, 19,636 acres. Population, 5,174.

BOYANAGH (or BOYOUNACH.)—Is a parish of County Galway, Connaught, 63 miles East Northeast of Danamore. Area, 15,832 acres. Population, 5,288.

BOYLE.—A barony of County Roscommon, Connaught, and recently divided in two parts—Boyle on the North and French Park on the South. Area, 94,283 acres; of which 65,137 are arable. There is Sandstone and Coal found in the North, and Sandstone in the South. There is also a town and parish in this barony situated on both sides of the river Boyle, and connected by a bridge 8 miles Northwest of Carrick-on-Shannon. Area of parish, 20,737 acres. Population, 12,591 ; of town, 3,235. Coarse Woolens are largely manufactured here. The Boyle River rises in Lough Gara, flows through Lough Key and minor Lakes, joining the Shannon river 1 mile Northwest of Carrick after a course of 13 miles.

BOYNE.—Is a river of Counties, Kildare, Kings, Meath and Louth, Leinster. It rises in the Bog of Allen near Carberry and empties into the Irish Sea about 4 miles below Drogheda. It is navigable for barges of 70 tons to Navan, 19 miles from the sea ; and at high water, for vessels of 200 tons to Drogheda, about 2¾ miles west of the last named town. A monument marks the spot where the forces of William III. on the first of July 1690, gained the victory over those of James II. so well known in British History, as the "Battle of the Boyne."

BRANDON.—Is a mountain, headland, bay and village of County Kerry, Munster. The mountain is 22 miles West of Tralee, is 3,126 feet in elevation and terminates Northeast in the headland, which forms the western limit of Brandon Bay, an arm of Tralee Bay. The village is on the West side of Brandon Bay, 10 miles Northeast of Dingle.

BRAY.—Is a maritime town and parish of Counties Dublin and

Wicklow, on the river Bray, 12 miles South Southeast of Dublin. Area of parish, 2,986 acres. Population, 3,326; of town, 3,169. The town is neatly built on both sides of the river which is here crossed by a bridge. It is a favorite summer resort for Sea-bathing.

BRAYHEAD.—Is a promontory about 1½ miles Southwest of the town, it is 807 feet above the sea.

BREAFY.—(or BREAGHWER.) Is a parish of County Mayo, Connaught, 2¾ miles East Southeast of Castlebar. Area, 5,266 acres. Population, 2,452.

BRIDE.—There are two rivers by this name. One rises in the Nagle Mountains, Counties, Cork and Wickford, Munster, and after an eastward course of 25 miles joins the Blackwater river 8 miles north of Youghal. On it are the towns of Rathcormarck and Tallow. It is navigable for barges to Kintaloon. The other in County Cork, Munster, after a course of 11 miles, joins the Lee, 6 miles West of Cork.

BRIDES BAY, (St.)—Is an inlet of the Irish Sea in the west extremity of County Pembroke. Ramsey and Skomer island are at its entrance and St. Davids and St. Brides are the principal towns on its shore.

BRIDGETOWN.—Is a parish of County Cork, Munster; it is 1 mile South of Castletown-Rothe. Area, 3,240 acres. Population, 993.

BRIGHT.—Is a parish of County Down, Ulster, 3 miles South Southeast of Downpatrick. Area, 5,334 acres. Population, 1,886.

BRIGOWN.—Is a parish of County Cork, Munster. Area, 15,221 acres. Population, 10,619. It comprises the town of Mitchelstown.

BROADHAVEN.—Is a bay on the West coast of County Mayo, Connaught, between Benwee and Errishead and 11 miles Northwest of Bangor. It is about 4 miles long.

BROSNA.—Is a parish of County Kerry, Munster; it is 8 miles East Southeast of Listowel. Area, 11,960 acres. Population, 2,871. The GREAT and LITTLE BROSNA are two small rivers flowing into the Shannon, County Kings, Leinster.

BROUGHSHANE.—Is a town of County Antrim, Ulster; it is 3½ miles East Northeast of Ballymena. Population, 940. Fairs, June 17 and September 1.

BROWNSTOWN.—Is a parish of County Meath, Leinster; it

is 5 miles South Southwest of Slane.   Area, 1,199 acres.   Population, 421.   There is a quantity of Copper Ore found in this parish.

BRUFF,—Is a town and parish of County Limerick, Munster ; it is 14½ miles South Southeast of Limerick. Area of parish, 1,331 acres. Population, 2,900 ; of town 1.393. It has a fine Church, large Roman Catholic Chapel and ruins of an old Castle. There are four Fairs held here Annually.

BRUREE.—Is a parish of County Limerick, Munster, 4 miles Northwest of Kilmalloch. Area, 3,210 acres. Population, 3,804, of which 703 are in the village. The Irish bards here held their half-yearly meetings untill 1746. There are four Fairs held here annually.

BUMLIN.—Is a parish of County Roscommon, Connaught. Area, 6,582 acres. Population, 5,257. It comprises a large part of Strokestown.

BUNCRAUA.—Is a market town of County Donegal, Ulster ; it is situated on Lough Swilly, 11 miles North Northwest of Londonderry. Population, 961. It has a handsome Church and Barracks and is much resorted to for Sea-bathing. Its Castle was an old seat of the O'Donnells.

BUNDORAU.—Is a maritime village and the principal watering place on the Northwest coast of Ireland, County Donegal, Ulster, on Donegal Bay, 4 miles Southwest of Balley-Shannon. Population, 299. BUNDROES is another village about 1 mile West Southwest.

BUNRATTY.—(Upper and Lower,) Are two baronies ; the Upper in County Clare, Munster, on Shannon river. The Lower in County Clare, Munster, 11 miles South Southeast of Clare. Area, 2,755 acres. Population, 1,320. It also contains a village, and an ancient Norman Castle of the 13th Century.

BUOLICK.—Is a parish of County Tipperary, Munster, 7 miles Southeast of Thurles. Area, 7,116 acres. Population, 2,660.

BURGESS.—(or BURGESSBEG) Is a parish of Tipperary, Munster, 5¼ miles South Southwest of Nenagh. Area, 4,980 acres. Population, 2,782.

BURRISHOOLE.—Is a maritime parish of County Mayo, Connaught. Area, 55,240 acres. Population, 11,942. It comprises the town of Newport.

BURT.—(BERT, or BIRT.) Is a parish of County Donegal, Ulster, on Lough Swilly, 6 miles West Northwest of Londonderry. Area, 10,673 acres. Population, 3,857.

BUSHMILLS.—Is a small town of County Antrim, Ulster, on the river Bush, 8 miles Northeast of Coleraine. Population, 788. It is a very neat town.

BUTTEVANT.—Is a market town and parish of County Cork, Munster. The town is on the river Arobeg, 3½ miles West of Doncraile. Area of parish, 11,583 acres. Population, 1,042; of town, 1,524. It was formerly enclosed by walls and it has the ruins of numerous ecclesiastical edifices, an old Castle and large Infantry Barracks. Fairs, March 27 and October 14.

CABLE ISLAND.—Is a small Island in County Cork, Munster, on the Atlantic Ocean, 5 miles South Southwest of Youghal.

CAHIR.—(or Caher.) Is a parish and market town of County Tipperary, Munster. The town is on the river Suir, 97 miles South Southwest of Dublin. Population, 3,668. It is very neatly built and has a handsome Church and Roman Catholic Chapel, &c., and near the town are the ruins of an Abbey and an old Castle in good repair with a Park adjacent. Friday is market day. Fairs, May 26 and 27, July 20, September 18 and 19 and December 7. There are also annual races. Area of the parish is 13,647 acres. Population of same, 8,801. There are, besides the above, two parishes of same name, as follows: County Kerry, Munster. Area, 19,110 acres. Population, 6,315. It comprises the town of Cahirciveen, which was the birth place of the late Daniel O'Connell. County Queens, Leinster 1¾ miles East Northeast of Borris-in-Ossory. Area, 1,826 acres. Population, 519. Also, two Islands, as follows: barony of Murrisk, County Mayo, Connaught, 4½ miles from the shore, and one in County Mayo, Connaught, 4½ miles South of Clare Island.

CAHIRAGH.—Is a parish of County Cork, Munster, 5 miles North of Skibbereen. Area, 23,516 acres. Population, 8,375.

CAHIRCIVEEN.—(or Cahirsiveen). Is a town of County Kerry, Munster, 2½ miles East Northeast of Valentia. Population, 1,492. It is of recent origin and has some public buildings.

CAHIRCONLISH.—Is a parish of County Limerick. Munster; it is 9 miles East Southeast of Limerick. Area, 8,173 acres. Population, 3,925. The Shannon Railroad passes within 2 miles of the village which contains 562 inhabitants.

CAHIRCONREE.—Is a mountain in County Kerry, Munster,

1,506 acres and a population of 1,052. The Lighthouse is on an abrupt cliff, 455 feet above the Sea. Latitude, 51°, 26′ North. Longitude, 9°, 29′ West. There is a ruined Castle and Church on the island, also a pier.

CAPPAGH.—There are two parishes by this name. One in County Tyrone, Ulster, 5 miles North Northeast of Omagh. Area, 37,671 acres. Population, 13,330. The splendid demesne of Mountjoy forest is here. Another in County Limerick, Munster, 2¾ miles North Northeast of Rathkeale. Area, 1,268 acres. Population, 755. There is a Copper Mine at Cappagh Hill, 10 miles West of Skibbereen. There is also a small river and bog of County Galway, Connaught, which has the same name.

CAPPAGHWHITE.—Is a town of County Tipperary, Munster; it is 7 miles North of Tipperary. Population, 1,046. There are five Fairs held here annually. There are also Copper Mines in this vicinity.

CAPPANACUSHY.—Are a group of islets at the head of the estuary and 3 miles West of Kenmore, County Kerry, Munster. On the mainland opposite, are the remains of Cappernacushy Castle.

CAPPOQUIN.—Is a town of County Waterford, Munster, on the river Blackwater, 4 miles East Northeast of Lismore. Population, 2,343. Cappoquin House is the seat of Sir R. Kean, Baronet.

CARDANGAN.—Is a parish of County Tipperary, Munster. Area, 3,906 acres. Population, 3,088. It comprises part of the town of Tipperary.

CARLINGFORD.—(or Carlinford) Is a petty maritime town and parish of County Louth, Leinster, on the South shore of Carlingford Bay, 10 miles East Northeast of Dundalk. Area, 24,050 acres. Population, 12,558; of town, 1,110. It is a miserable collection of cabins, whose inhabitants are mostly engaged in fishing. Markets, Tuesday and Saturday. Fair, October 10. CARLINGFORD BAY, is an inlet of the Irish Sea, between Counties, Louth and Down; it is 11 miles in length and connected to Lough Neagh by the Newry canal. There is a Lighthouse on an island at its mouth, Latitude, 54°, 1′ North. Longitude, 6°, 5′ West. Carlingford mountains rise on the South side of the Bay, to the height of 1,935 feet.

CARLOW.—Is an inland County of Leinster. It is encircled

by Counties, Kildare, Wicklow, Wexford and Kilkenny. Area, 221,342 acres, of which 185,000 acres are arable, and 31.000 acres are bog, mountain and waste land. Population, in 1851, 68,073. The surface is flat or generally undulating. The Barrow and Slaney are the principal rivers. Agriculture is very much advanced, and there are numerous Dairy farms. Average rent of land 15 shillings per acre, bringing in an aggregate rental of about £150,000 per annum. There are no important manufactures, but granite abounds throughout the County. The principal exports are Corn, Flour, Malt, Barley and Butter. There are 6 baronies and 50 parishes in the diocese of Leighlin. The chief towns are Carlow and Old Leighlin. It sends two members to the House of Commons exclusive of its Capitol.

CARLOW.—Is a parliamentary and municipal borough, town and parish of County Carlow, Leinster, at the confluence of the Barren and Barrow rivers, 44 miles South Southwest of Dublin, to which it communicates by Railroad. Area of parish, 3,330 acres. Population, 9,901. Area of parliamentary borough, 572 acres. Population in 1851, 10,955. It is clean and well built and contains handsome buildings, both public and private, it also has two handsome bridges. It exports Corn, Bacon and excellent Butter to Waterford by the river, and to Dublin by the Grand canal. Markets for Produce, Mondays and Thursdays. Fairs, May 4, June 22, August 26 and November 8. County Assizes and Quarter and Petty Sessions are held, and a County Police force is stationed at Carlow. It sends one member to the House of Commons.

CARMEN.—Is a township of County Kildare, Leinster, 9 miles East of Athy. There are druidical remains here.

CARN.—Is a small market town of County Donegal, Ulster, 16 miles North of Londonderry. Population, 653.

CARNE.—(or CARNA) Are two parishes of Leinster. One in County Wexford, 3 miles South Southeast of Broadway. Area, 1,963 acres. Population, 919. Another in County Kildare, 4 miles East Southeast of Kildare. Area, 1,157 acres. Population, 499. There is a hamlet by this name in County Mayo, Connaught,

CARNEW.—Is a township and parish of County Wicklow, Leinster, 7 miles West of Gorey. Area of parish, 23,446 acres. Population, 7,205; of town, 979. The town is well built and contains the remains of a Castle. There are five Fairs held here

Annually. Near it is Coolatlin, the mansion of Earl Fitzwilliam, chief proprietor of this parish.

CARNMONEY.—Is a parish of County Antrim, Ulster, 6 miles North of Belfast. Area, 8,937 acres. Population, 6,128.

CARNSORE POINT.—Is a headland forming the Southeast extremity of the Irish mainland, County Wexford, Leinster, 12 miles South Southeast of Wexford.

CARRA.—Is a barony of County Mayo, Connaught, running from the North Northeast to South Southwest. On its North boundary are Loughs, Cullen and Con; and on its South, Loughs, Carra and Mask; on the Southeast the ground is low and cultivated; and on the North it is mountainous and moorland. The scenery is very picturesque.

CARRAN-TUAL.—Is the highest mountain of Ireland. It is in County Kerry, Munster, in the range of the MacGillicuddy Reeks, 5 miles Southwest of Killarney. Height, 3,414 feet.

CARRENTED.—Is a parish of County Tyrone, Ulster. Area, 13,432 acres. Population, 7,903. It includes the town of Aughnacloy.

CARRICK.—There are several parishes by this name in the Province of Leinster, as follows, viz: County Wexford, 2½ miles West of Wexford; Area, 3,009 acres. Population, 1251; County Kildare; Area, 5,196 acres. Population, 552; County Westmeath; Area, 2,957 acres. Population, 532; (or CARRICKBAGGET,) County Louth; Area, 826 acres. Population, 302.

CARRICK.—(or CARRICKAHOOLAY) Is an old Tower of County Mayo, Connaught, 5 miles West Northwest of Newport.

CARRICK.—(or CARRICK-A-REDE) Is a Rock of County Antrim, Ulster, 2 miles West of Kenbane Head.

CARRICKBEG.—(formerly CARRICKMACGRIFFIN) Is a town of County Waterford, Munster, 14 miles West Northwest of Waterford on the river Suir, here crossed by a bridge which connects it with Carrick-on-Suir. Population, 2,680. It contains the remains of a Castle of the 14th Century.

CARRICKFERGUS.—Is a parliamentary and municipal borough, market town and parish of County Antrim, Ulster, on Belfast Lough, 9 miles North Northeast of Belfast. Area of parish, 16,700 acres. Population, 8,488. The town which was formerly a stronghold has the remains of fortifications built in 1576. The houses are mostly stone, but the streets are dull and dirty.

The parish Church having noble monuments, including those of the Chichester family, communicated formerly by a still existing subteranean passage with a Monastery, on the site of which Sir Arthur Chichester erected the noble Castle of Joymount. Carrickfergus Castle, built about 1128, is still used as an arsenal: there are other fine edifices. They have a good pier for vessels of 100 tons, but trade is not good. There are some small manufactories of Linen and Cotton Fabrics here. Markets, Wednesday and Saturday. Fairs, May 12 and November 1. This borough sends one member to the House of Commons. William the III. landed here in 1690. In 1760 the town and Castle was taken by the French, who were soon driven away.

CARRICKMACROSS.—Is a market town and parish of County Monaghan, Ulster, on the road from Dublin to Londonderry, 12 miles Southwest of Dundalk. Area, 16,702 acres. Population, 13,444; of town, 1,997. It is well built and has the largest Distillery in the district. There are five Fairs held here Annually. The ruins of a Castle built by the Earl of Essex, to whom the town was granted by Queen Elizabeth, and in the possession of whose family the estate remains, is in this town.

CARRICK-ON-SHANNON.—Is a disfranchised borough and market town of County Leitrim, Connaught, on the river Shannon, 19 miles North Northwest of Longford. Population, 1,984. It is ill-paved but has some handsome buildings. There are three Fairs held here Annually.

CARRICK-ON-SUIR.—Is a market town and parish of County Tipperary, Munster, on the river Suir, which is here crossed by a bridge built in the 14th Century, it is 13 miles South of Clonmel. Area of parish, 2,426 acres. Population, 9,165; of town, 8,831. It was formerly enclosed by walls and contains a very antique Church, a fine Roman Catholic Chapel and a Castle formerly belonging to the Ormonde family. Recent improvements in the river enable vessels of large tonnage to approach the town, which has an export trade in Corn and Cotton. Fairs, monthly. The country is fertile and well wooded. Curraghmore, the seat of the Marquis of Waterford, is about 4 miles South.

CARRIGALINE.—(or BEAVER) Is a maritime parish of County Cork, Munster, 8 miles Southeast of Cork harbor. Area, 14,498 acres. Population, 7,489; mostly employed in the Marble and Slate Quarries. The village, which is now unimportant, bid fair once to be the rival of Cork through the instrumentality of the Earl of Cork. The picturesque ruins of Carrigaline, a Castle of the Desmonds, are here, also the remains of a religious house and a Danish Fort. Fairs, Easter Monday, Whitsunday, August 12 and November 8.

CARRIGALLEN.—Is a barony of County Leitrim, Connaught. Area, 63,501 acres. It is drained by an affluent of the river Erne. There is a parish of same name in County Leitrim, Connaught, 11 miles East Northeast of Mohill. Area, 18,104 acres. Population, 8,100. There is some beautiful scenery in the vicinity.

CARRIGDOWNAM.—Is a parish of County Cork, Munster. Area, 797 acres. Population, 245.

CARRIGNAVAR.—(or DUNBOLLOGE) Is a parish of County Cork, Munster, 5 miles North of Cork. Area, 16,783 acres. Population, 5,269.

CARRIG-O-GUNNELL.—(or CARRICKAQUICY) Is a village of County Limerick, Munster, 5 miles West Southwest of Limerick. Its ruined Castle on a lofty rock was formerly the stronghold of the O'Briens, Kings of Munster, but blown up at the siege of Limerick in 1691.

CARRIGROHANE.—Is a parish of County Cork, Munster, on the river Lee, 2 miles West of Cork. Area, 2,658 acres. Population, 2,279. There are ruins of several Castles on this river.

CARRIGTOHILL.—Is a parish of County Cork, Munster, 3 miles West of Middleton. Area, 10,319 acres. Population, 3,976; of village, 692. It has an ancient Church, and in various parts of the parish are curious subteraneous chambers within circular intrenchments, called Danish Camps. There are five Fairs held here Annually.

CARRIGAFOYLE.—Is a small island in the estuary of the Shannon river, County Kerry, Munster, 2 miles North of Ballylongford. It has a Castle, once the seat of the O'Connor-Kerry.

CARROWMORE.—Is a lake of County Mayo, Connaught, 4 miles Northeast of Tulloghan Bay. It contains several islets and discharges itself by the river Munhin into the Owenmore.

CARYSFORT.—(MACREDDIN or MOYENEDIN) Is a disfranchised borough of County Wicklow, Leinster, 5 miles Southwest of Rathdrum. It gives the title of Earl to the Proby family, in whose fine seat near the village is a curious ancient obelisk; elevation 100 feet.

CASHEEN BAY.—Is a bay in County Galway, Connaught, on the West side of the island Caromma; it is easy of access, and has a depth of water for large ships.

CASHEL.—Is a city and parliamentary and municipal borough of County Tipperary, Munster, 49 miles North Northeast of Cork, on the road to Dublin. Population of city in 1851, 4,793; of borough, 8,027. It is situated in the centre of a fine agricultural country at the foot of the Rock of Cashel, a limestone height, on the top of which is the most interesting assemblage of ruins in Ireland. The town is miserably built, with the exception of the principal streets, on which are some fine buildings, both public and private. Near the town is the remains of Hore Abbey and of a Dominican Priory. This is an Archbishop's See, now combined with the See of Waterford, where the diocesan resides. Markets, Wednesday and Saturday. Fairs, March 26, August 7 and on third Tuesday of each month. Donald O'Brien, King of Limerick, and his nobles, swore allegiance to Henry II. at Cashel in 1172. It is also the name of a parish of County Longford, Leinster, 5 miles South of Lanesborough. Area, including Loughs, 22,151 acres. Population, 5,559.

CASTLEBAR.—(or AGLISH) Is a disfranchised parliamentary and municipal borough, town and parish of County Mayo, Connaught, on the river Castlebar, 10 miles East Northeast of Westport. Area of parish, 14,974 acres. Population, 10,464; of town, 5,137. It is a poorly built town and stands on a plain of bog and pasture land. It has the usual buildings, a Church, Roman Catholic Chapel, Court House, etc., and Barracks for 650 men. It has some Breweries and a considerable trade in Coarse Linens and rural products. It is also the head of the Poor-Law Union and the seat of the County Court of Assize. Market on Saturday. Fairs, May 11, June 19, September 16 and November 18. In the immediate vicinity are "the Park" and "the Lawn," seats of the Earl of Lucan and of St. Clair O'Malley, Esq. It was taken in 1798 by the French under General Humbert, who in an action derisively called the "Race of Castlebar" defeated a superior British force here. The Castlebar River issues from a lake 3 miles in length, Southwest of the town, and flows North into Lough Cullin.

CASTLE BLAKENEY.—(or KILLASOLAN) Is a parish of County Galway, Connaught, 18 miles Southeast of Tuam. Area, 11,483 acres. Population, 4,496. The country is poor and boggy and the village is miserable. There are five Fairs held here Annually.

CASTLE BLAYNEY.—Is a town of County Monaghan, Ulster, 12 miles South Southeast of Monaghan, at the West of Lough Blayney. Population, 2,134. It is well built and has a Church, Market House, Work House, etc. Markets on Wednesday. Fairs, first Wednesday in every month. It gives the title of Viscount to the Blayney family whose desmene is in the vicinity.

CASTLE COMER.—Is a town and parish of County Kilkenny, Leinster, 10 miles North Northeast of Kilkenny on the road to Dublin. Area of parish, 21,592 acres. Population, 13,585; of town, 1,765. It is regularly built and clean, and stands in a hollow. It has a large Church and various Schools, etc. Market, Saturday for dairy and field produce. There are six Annual Fairs. There are extensive Collieries 2½ miles distant. Castle Comer and ruins are also in the vicinity.

CASTLE CONNEL.—(or STRADBALLY) Is a town and parish of County Limerick, Munster, on the river Shannon, near the Falls of Doonass, 6½ miles North Northeast of Limerick. Area of parish, 6,698 acres. Population, 5,433; of town, 1,106. It is beautifully situated and is neat and clean, and is much resorted to in summer by the people of Limerick for its Chalybeate Springs. Its Castle, formerly a seat of the O'Briens, Kings of Munster, was destroyed during the siege of Limerick.

CASTLE CONNER.—Is a parish of County Sligo, Connaught, on the river Moy, 3 miles North Northeast of Ballina. Area, 16,678 acres. Population, 5,136. The ruins of an old Castle which gives name to the parish, is here.

CASTLE DERMOT.—(or THISTLEDERMOT) Is a parish and ancient town of County Kildare, Leinster, on the river Lear, 6 miles North Northeast of Carlow. Area of parish, 7,498 acres. Population, 3,090; of town, 1,516. Its numerous antiquities comprise remains of a large Cathedral, of a Church built by the first English settlers, of a beautiful Franciscan Monastery, a Norman Arch, a strong square Tower supposed to have been built by the Knights Templars, the ruins of a Priory and a Castle; and in its Churchyard are several curious crosses and a Round Tower. There are six Fairs held here Annually. The town was formerly the residence of the Dermots, Kings of Leinster.

CASTLE HAVEN.—Is a parish of County Cork, Munster, on

Castle Haven Bay, 15 miles Northeast of Cape Clear. Area, 10,542 acres. Population, 6,056.

CASTLE ISLAND.—Is a town and parish of County Kerry, Munster, 11 miles East Southeast of Tralee. Area of parish, 29,633 acres. Population, 7,967; of town, 1,687. It has some fine buildings. There is an islet of same name in County Cork, Munster, in Roaring-water Bay, North of Cape Clear. Area, 120 acres.

CASTLE JORDAN.—Is a parish of Counties, Kings and Meath, Leinster, on the Grand Canal, 5 miles South Southwest of Kinnegad. Area, 17,372 acres. Population, 4,079.

CASTLE KNOCK. Is a parish of County Dublin, Leinster, in a barony of the same name and 4 miles West Northwest of Dublin on the river Liffey. Area, 7,124 acres. Population, 4,063. It has a new Church and the ruins of a Castle built in the reign of Henry II. and those of an Abbey built in the 13th Century, endowed with £800 per annum.

CASTLE LYONS.—Is a village and parish of County Cork, Munster, 2 miles Northeast of Rathcormack. Area of parish, 12,710 acres. Population, 5,526; of village, 775. It has a Carmelite Monastery and ruins of a Dominican Priory.

CASTLE MACADAM.—Is a parish of County Wicklow, Leinster, in the vale of Ovoca, 6 miles South of Rathdrum. Area, 10,843 acres. Population, 5,633.

CASTLE MAGNER.—Is a parish of County Cork, Munster, 7 miles West Northwest of Mallow. Area, 7,880 acres. Population, 3,007. Its antiquities include an ancient baronial Castle of the Magner's. It comprises part of the town of Kanturk.

CASTLE MARTYR.—Is a small town and formerly a parliamentary borough of County Cork, Munster, 18 miles East of Cork, on the river Maine, by which it has a good trade. Population, 1,397. The Earl of Shannon is proprietor of the adjacent demesne. There are four Fairs held here Annually.

CASTLE MORE.—Is a parish of Counties, Roscommon and Mayo, Connaught, 1 mile Southeast of Ballaghadireen. Area, 8,914 acres. Population, 3,582.

CASTLE POLLARD.—Is a town of County Westmeath, Leinster, 6½ miles West Northwest of Drumcree. Population, 1,310. It is well built, clean, and superior in its general looks to most towns of its size. It has an elegant Church. About 1 mile West is Pakenham Hall, seat of the Earl of Longford.

CASTLE RAHAN.—Is a parish of County Cavan, Ulster, 5 miles West of Virginia. Area, 10,315 acres. Population, 7,589. It comprises part of the town of Ballyjamesduff.

CASTLE-REA.—(or CASTLEREAGH) Is a market town in barony of same name, County Roscommon, Connaught, 16 miles West Northwest of Roscommon, on the river Suck, here crossed by two bridges. Population, 1,255. It consists chiefly of one long street. Castlerea Hall is the property of Lord Mountsandford, on whose demesne are the ruins of the ancient Castle. CASTLEREAGH, which gives the title of Viscount to the Marquis of Londonderry, is a hamlet of County Down, Ulster, 2 miles Southeast of Ballyacerret.

CASTLE TERRA.—Is a parish of County Cavan, Ulster, 4 miles North Northeast of Cavan. Area, 9,981 acres. Population, 6,813.

CASTLETOWN.—Is the name of a seaport town and several parishes of Ireland. The town is in County Cork, Munster, on the West side of Bantry Bay, 18 miles West of Bantry. Population, 861. Vessels of 400 tons can reach its pier. There are nine Fairs held here Annually. The parishes are situated as follows : County Louth, Leinster, 1½ miles North Northwest of Dundalk ; Area, 2,611 acres. Population, 1,043 ; County Limerick, Munster, 4 miles Northeast of Pallas Green ; Area, 1,777 acres. Population, 919 ; County Tipperary, Munster, 7 miles Northwest of Nenagh ; Area, 9,274 acres. Population, 4,292 ; County Meath, Leinster, 7 miles West of Athboy ; Area, 12,282 acres. Population, 4,588. Near the village is Clonyn Castle, the seat of the Marquis of Westmeath.

CASTLETOWN ROCHE.—Is a town and parish of County Cork, Munster, on the Awbeg, 8 miles West Northwest of Fermoy. Area of parish, 6,485 acres. Population, 3,476 ; of town, 1,063. It stands on a wooded height. There are four Fairs held here Annually.

CASTLETOWNSEND.—Is a small seaport town of County Cork, Munster, on the West side of Castlehaven Bay, 4 miles East Southeast of Skibbereen. Population, 770. It is the Custom House for the port of Baltimore.

CASTLE WELLAN.—Is a market town of County Down, Ulster, 5 miles Southwest of Clough. Population, 806. It gives the title of baron to the Earl of Anuesley, Lord of the manor, whose seat, Castle Wellan, adjoins the town.

CASTROPETRE.—Is a parish of County Kings, Leinster, 8 miles South Southeast of Kinnegad. Area, 15,762 acres. Population, 432. It comprises part of the town of Edenderry.

CAVAN.—Is an inland County of Ulster, surrounded on the North by County Fermanagh, on the South by County Meath, on the East by County Monaghan and on the West by County Meath. It is touched on the Southwest by County Leitrim. Area, 477,360 acres, of which 275,473 acres are arable and the balance, (with the exception of 22,142 acres which is water,) is waste land. Population, 1851, 174,303. The Country is open and good and is enclosed on its borders by ranges of mountains. The Woodford and Upper Erne are the principal rivers. Chief Loughs, are Gawnagh, Shillen, etc. The scenery on some of these is highly picturesque. Agriculture is very backward, as the soil, with the exception of the banks of the rivers, is light and poor. The principal manufacture is Linen. Cavan is sub-divided into 8 baronies and 36 parishes and sends 2 members to the House of Commons. Cavan and Belturbet are the principal towns. CAVAN, a market town, (formerly a parliamentary borough,) is the Capital and is situated on a branch of the river Annalee, 26 miles South Southeast of Enniskillen, on the Railroad to Dublin. Population, 3,740. It is very poorly built but contains some very handsome buildings. The seat of Lord Farnham is immediately adjacent. Markets on Tuesday. There are seven Fairs held here annually. It is also the head of the Poor Law Union.

CELBRIDGE.—Is a town and parish of County Killdare, Leinster, on the river Liffey, 15 miles Southwest of Dublin. Population, 1.289. It is well built and contains a Church, Work House, County Hospital and a large Woolen Factory. Fairs last Tuesday in April, September 8 and November 7. In the vicinity are Killadoon, the seat of the Earl of Leitrim, and Lyons, the seat of the Earl of Cloncurry. It is also the head of the Poor Law Union.

CHAPEL IZOD.—Is a town of County Dublin, Leinster, 3 miles west of Dublin, situated on the river Liffey. Population, 1,575. It contains an ancient Church and Barracks.

CHARLEMONT.—Is a market town of County Armagh, Ulster, on the river Blackwater, 6 miles North Northwest of Armagh. Population, 485. There is a strong fort used as the ordnance depot and head Artillery quarters of North Ireland. It gives the title of Earl to the Caufield family.

CHARLEVILLE.—(or RATHGOGAN) Is a town, parish and municipal borough of County Cork, Munster, 22 miles West Southwest of Limerick. It is built very good and consists of four streets crossing each other at right angles. It gives the title of Earl to the Barry family ; but the town and vicinity belong to the Earl of Cork and Ossory, whose mansion here was burnt by the Duke of Berwick in 1690.

CHURCHTOWN.—(or BRUHENNY) There are several parishes by this name, as follows : County Cork, Munster, 7 miles South Southwest of Charleville ; Area, 8,047 acres. Population, 3,777. The village is neatly built and near it are the seats of Burton and Eymont, giving the title of Earl to the Percival family. (or RHEBAN) is in County Kildare, Leinster. Area, 7,331 acres. Population, 2,294. It is traversed by the Grand Canal ; County Westmeath, Leinster, 5 miles West Southwest of Mullingar ; Area, 5,302 acres. Population, 1,108 ; County Meath, Leinster, 3 miles Southwest of Navan. Area, 1,336. Population, 509. There are also villages of the same name in Counties, Waterford, Limerick and Wexford.

CLAGGAN BAY.—Is a bay on the coast of County Galway, Connaught, 3 miles West Northwest of Innisboffin. It is a safe harbor for vessels of the largest tonnage, and has a good dock.

CLANE.—Is a barony, parish and town of County Kildare, Leinster. The town is on the river Liffey, 7 miles West Southwest of Selbridge. Area of parish, 4,663 acres. Population, 2,160 ; of town, 335. The remains of an Abbey founded in 548, and of a Franciscan Priory of the 13th Century, are still to be seen. There are four Fairs held here Annually Adjoining it is the Bog of Clane. Area, 2,235 acres.

CLANMAURICE.—Is a barony of County Kerry, Munster, bounded on the West by the Atlantic Ocean. It is 17 miles in length.

CLANMORRIS.—Is a barony of County Mayo, Connaught, 18 miles in length, from North to South.

CLANWILLIAM.—Is a barony of County Limerick, Munster, having the Shannon river on its North, and is 10 miles in length. Is also the name of a barony of County Tipperary, Munster, bounded on the West by County Limerick, 18 miles in length; forming a fertile and picturesque district.

CLARA.—Is a market town and parish of County Kings, Leinster, 5 miles Southwest of Kilbeggan, near the Brosna river. Population, 1,155. There are eight Fairs held here Annually. County Sessions and a Market weekly for Corn. Is also the name of a parish of County Kilkenny, Leinster, 3 miles East Northeast of Kilkenny. Area, 3,201 acres. Population, 663.

CLARA.—Is an island off the West coast of County Mayo, Connaught, at the entrance to Clew Bay, 4 miles South of Achil island. Its Lighthouse is on the North point at an elevation of 487 feet. Latitude, 53° North. Longitude, 9° 59′ West. Area, 3,959 acres. Population, 1,616. Its surface is mountainous, its highest point being 1,520 feet above the sea.

CLARE.—Is a maritime County of Munster, situated with the Atlantic on the West, and on the East, North and South the Counties, Galway, Tipperary and Limerick, and being separated from the two latter by Lough Derg and the Shannon river. Area, 827,994 acres; of which 445,009 acres are arable, 8,384 in plantations, and the balance being either water or waste. Population in 1851, 212,428; occupying 44,870 houses. The surface is mostly hilly and rugged, but there is some fine level land. The Fergus and its branches are the principal rivers. Average rent of land, 11s. 3d. per acre. The fisheries are very important, but there are no manufactories of any account. Clare is subdivided into 11 baronies and 80 parishes, in the dioceses of Kilfenora, Killaloe and Limerick. The chief towns are Ennis, (the capital) Kilrush, Ennistimon and a part of Killaloe. This County sends 2 members to the House of Commons. There are reputed to be the remains of 118 baronial Castles in this County.

CLARE.—Is a town of County Clare, Munster, on the river Fergus, 2 miles East Southeast of Ennis. Population, 879; of parish of Clare Abbey, 3,280. It is pleasantly situated and regularly built and clean, and it is the port for all the exports of the centre of the County, but has a very poor dock; totally inadequate to its wants. Fairs, June 3 and November 11. It gives the title of Earl to the Fitzgibbon family. 1 mile South, is the remains of an Abbey founded by O'Brien, King of Munster, in the 13th century. (CLARE-MORRIS,) is a town of County Mayo, Connaught, 15 miles Southeast of Castlebar. It is clean and neat. Population, 2,236.

CLARE.—Is a river of County Galway, Connaught, 32 miles long. Its course is generally Southward and it flows into Lough Corrib, 3 miles North of Galway. In many places the river is very shallow and marshy and for 3 miles it is subterraneous. The towns of Dunmore and Clare are situated on its banks.

CLARE.—Is a barony of County Galway, Connaught, separated from County Mayo by the Black river. Length, 19 miles.

CLARE-GALWAY.—Is a parish of County Galway, Connaught, 6 miles North Northeast of Galway. Area, 12,453 acres. Population, 4,042. The surface is fertile, being drained by the Clare river.

CLASHMORE.—Is a village and parish of County Waterford, Munster, 4 miles North Northeast of Youghall. Area, 7,202 acres. Population, 3,777. It has a County General Sessions. There are four Fairs held here Annually.

CLEENISH.—Is a parish of County Fermanagh, Ulster, 7 miles West Southwest of Enniskillen. Area, 36,681 acres. Population, 11,075. In it are Loughs Erne and Macnean, and on an island in Lough Erne are the remains of an Abbey, now used as a parish Church.

CLEW BAY.—Is an inlet of the Atlantic Ocean, of County Mayo, Connaught. Latitude, 53° 55′ North. Longitude, 9° 50′ West, extending inland for about 15 miles, with a uniform breadth of 8 miles. There are numerous small harbors and fishing stations on the shores, among which are the towns of Newport, Westport and Louisburgh. At its upper end is an Archipelago of about 300 fertile and cultivated islets. Clare island is opposite its entrance.

CLIFDEN.—Is a seaport town of County Galway, Connaught, in the district of Connemara and 43 miles West Northwest of Galway, on an inlet of Ardbear harbor. Population, 1,509. It has fine public buildings and Churches, and exports considerable Oats and imports American Timber. Markets, weekly. There are seven Fairs held here Annually. Clifden Castle is close by. Is also the name of a village of County Kilkenny, Leinster, which gives the title of Viscount to the Agar-Ellis family.

CLOGHANE.—(or CLAHANE) Is a parish of County Kerry, Munster, West of Brandon Bay. Area, 17,572 acres. Population, 2,994. It consists mostly of a rocky peninsular.

CLOGHUN.—Is a market town of County Tipperary, Munster, 13 miles West Southwest of Clonmell. Population, 2,049. The seat of the Viscount Lismore, Shanbally, is within 2 miles of this town. The are four Fairs held here Annually.

CLOGHER.—Is a market town and parish of County Tyrone, Ulster, situated on the Blackwater river, 82 miles North Northwest of Dublin. Area, 49,761 acres. Population, 17,813 ; of town 702. It is in a valley and its surface is rich and undulating. The principal buildings of the town are the Cathedral, Bishop's Palace, Prison and Work-House. Fairs on the 20th of each month, also on May 2 and July 26. It was formerly an Episcopal town and parliamentary borough.

CLOGHER.—There are several parishes by this name, as follows, viz : County Tipperary, Munster, 6 miles Southwest of Thurles ; Area, 8,119 acres. Population, 2,643 ; (or KILCLOGER) Is a parish of County Louth, Leinster, 7 miles Northeast of Drogheda ; Area, 1,861 acres. Population, 1,371. Its village is much resorted to for Sea-bathing. There are six Fairs held here Annually. County Mayo, Connaught near Kilcumminghead, and 4 miles North of Killala. It was here that the French expedition landed in 1798.

CLOGHERNEY.—(or CLOUGHERNEY) Is a parish of County Tyrone, Ulster, 5 miles Southeast of Omagh. Area, 17,792 acres. Population, 7,553.

CLONAKILTY. (or CLOUGHNAKILTY) Is a market town of County Cork, Munster, on the Foilagh river, (here crossed by two bridges,) 11 miles Southwest of Bandon. Population, 3,993. It was formerly a flourishing parliamentary town. Linens to the value of £30,000 being woven here Annually ; but with the exception of some Cotton Manufactories the town is dead. There are five Fairs held here Annually.

CLONALLON.—Is a parish of County Down, Ulster, 1 mile North of Warrenpoint. Area, 11,658 acres. Population, 6,553.

CLONARD.—Is a parish of County Meath, Leinster, 15 miles West Northwest of Kilcock. Area, 13,324 acres. Population, 4,508. The village is on the river Boyne, which is here crossed by a bridge. The ruins of Turogan Castle are South of the village. The surface is flat and boggy.

CLONBEG.—Is a parish of County Tipperary, Munster, 4 miles Southwest of Tipperary. Area, 15,112 acres. Population, 4,377. Galtee-More, the highest point of the Galtee mountains, is in this parish. Height, 3,015 feet.

CLONBERN.—Is a parish of County Galway, Connaught, 5½ miles Southeast of Dunmore. Area, 10,462 acres. Population, 2,333. Loughs, Mackeeran and Doo are in this parish.

CLONBRONEY.—(or CLONEBRONE) Is a parish of County Longford, Connaught, 6 miles West of Granard. Area, 12,708 acres. Population, 5,114.

CLONBULLOGE.—There are two parishes by this name. Clonbulloger (or Clonsast) is a parish of County Kings, Leinster, 6 miles Southwest of Edenderry. Area, 23,558 acres. Population, 3,803. And another in County Tipperary, Munster, 5 miles Southeast of Tipperary. Area, 3,956 acres. Population, 1,546.

CLONCHA.—(or CLONCA) Is the most Northern parish of Ireland. County Londonderry, Ulster. It is situated between the Strabreasy Bay and the Atlantic Ocean. Area, 19,643 acres. Population 6,798. Malin-head and Well and various other antiquities are in this parish. The surface is mountainous.

CLONCLARE.—(or CLOONCLARE) Is a parish of County Leitrim, Connaught. Area, 32,900 acres. Population, 10,524. It comprises part of the town of Manor-Hamilton.

CLONCURRY.—There are two parishes of County Kildare, Leinster, by this name. One 5½ miles Northwest of Kilcock. Area, 8,390 acres. Population, 1,666. It gives the title of Baron to the Lawless family. Another, 3 miles East Northeast of Rathangan. Area, 5,240 acres. Population, 644. The surface is boggy and is crossed by the Grand Canal.

CLONDAGAD.—(or CLONDEGAD) Is a parish of County Clare, Munster, 6½ miles South Southwest of Clare. Area, 16,978 acres. Population, 5,088.

CLONDALKIN.—Is a parish of County Dublin, Leinster, 4½ miles West Southwest of Dublin. Area, 4,934 acres. Population, 2,496. The village has a School House, Alms House, etc.

CLONDEHORKEY.—(or CLONDAHORKEY) Is a parish of County Donegal, Ulster, 16 miles West Northwest of Letterkenny. Area, 29,633 acres. Population, 6,908. Murkish mountain in this parish is 2,190 feet in height. The surface is mountainous and poor.

CLONDERALAW BAY.—Is a bay of County Clare, Munster, 18 miles Southwest of Ennis, 4 miles in length and from ½ to 1½ miles in breadth.

CLONDERADOCK.—(or CLONDARADOG) Is a parish of County Donegal, Ulster, 9 miles North of Rathmullen. Area, 27,367 acres.

Population, 10,344. It occupies the greater part of the peninsular of Fannat.

CLONDROHOD.—Is a parish of County Cork, Munster, 3 miles Northwest of Macroom. Area, 27,114 acres. Population, 6,258. There are some intrenchments here said to have been built by the Danes.

CLONDUFF.—(or CLANDUFF) Is a parish of County Down, Ulster, 3 miles Southeast of Rathfriland. Area, 21,242 acres. Population, 8,687. Eagle mountain, 2,084 feet in height is in this parish.

CLONE.—There are two parishes by this name. CLONE (or CLOONE) is in County Leitrim, Connaught, 4 miles Northeast of Mohill. Area, 41.523 acres. Population, 21,225; and another in County Wexford, Leinster, 3 miles East Northeast of Enniscorthy. Area, 6,267 acres. Population, 1,504.

CLONENAGH and CLONAGHEEN,—Is a parish of County Queens, Leinster, 7 miles West Southwest of Maryborough. Area, 47.189 acres. Population, 18,403. It includes the town of Mountrath.

CLONES.—Is a parish and market town of County Monaghan, Ulster, 11 miles West Southwest of Monaghan and near the Ulster Canal. Area, 42,878 acres. Population, 23,506; of town, 2,877. The town contains various public buildings. Markets, weekly. Fairs, last Thursday of every month. Near it are numerous antiquities, also the ruins of an old Abbey founded in the 13th Century. It is also the head of the Poor Law Union.

CLONEY.—(CLONY or CLONIE) Is a parish of County Clare, Munster, 5 miles East Northeast of Ennis. Area, 10,656 acres. Population, 3,624.

CLONFANE.—Strawberry Hill and Queensfort, (Bog of,) County Galway, Connaught. The river Clare rises here. It comprises 3,715 acres.

CLONFEACLE.—(or CLUAIN FIACUL) Is a parish of Counties, Armagh and Tyrone, Ulster. Area, 26,218 acres. Population, 18,930. It comprises the town of Moy.

CLONFERT and KILMORE.—Are two contiguous bogs in County Galway, Connaught, situated Northwest of the Shannon river. They comprise 9,615 acres, with an average depth of 30 feet. They are traversed by the Grand Canal.

CONFERT.—There are two parishes by this name. CONFERT

(or NEW-MARKET) is in County Cork, Munster. Area, 62,110 acres. Population, 17,328. It comprises the town of New-market and part of Kanturk. And another in County Galway, Connaught, 4½ miles North Northeast of Eyrecourt. Area, 24,877 acres. Population, 5,704. Surface flat and boggy. It is traversed by the Grand Canal.

CLONFINLOUGH.—(or CLOONFINLOUGH) Is a parish of County Roscommon, Connaught, 3 miles South of Stokestown. Area, 7,814 acres. Population, 4,782.

CLONGESH.—(or CLONGISH) Is a parish of County Longford, Leinster, 3 miles North Northwest of Longford. Area, 12,833 acres. Population, 6,504.

CLONKEEN.—There are several parishes by this name, as follows, viz: County Louth, Leinster, 4 miles Northwest of Ardee. Area, 4,322 acres. Population, 2,158; (or CLONKEEN-KERRY) County Galway, Connaught, 7 miles Northeast of Athenry. Area, 8,214 acres. Population, 1,971; County Limerick, Munster, 5½ miles East of Limerick. Area, 1,145 acres. Population, 621.

CLONLEIGH.—There are several parishes by this name, as follows, viz: County Donegal, Ulster. Area, 12,517 acres. Population, 5,686. It comprises part of the town of Lifford; (or CLONLEE,) County Clare, Munster, 4 miles East Northeast of Six-mile-Bridge. Area, 8,834 acres. Population, 3,749; County Wexford, Leinster, 54 miles Northeast of New Ross. Area, 2,717 acres. Population, 830.

CLONMACNOISE.—(THE SEVEN CHURCHES) Is a parish of County Kings, Leinster, 7 miles South Southwest of Athlone. Area, 22,417 acres. Population, 4,775. The ruins of a Cathedral and various Monastic buildings are still to be seen in and around the village.

CLONMANY.—Is a parish of County Donegal, Ulster, 9 miles North Northeast of Buncrana. Area, 23,376 acres. Population, 6,489.

CLONMEEN.—(or CLOONMEEN) Is a parish of County Cork, Munster, 10 miles West of Mallow. Area, 20,076 acres. Population, 6,364.

CLONMEL.—Is a parliamentary and municipal borough and town of Counties, Waterford and Tipperary, Munster, on the banks of the river Suir, and the Limerick and Waterford Railroad, 14 miles South Southeast of Cashel. Area of parliamentary borough,

331 acres. Population, 12,386. Its streets are regularly built, paved and lighted with gas. It is a thriving town and contains many manufactories. Its principal buildings are a Church, Roman Catholic and other Chapels, Jail, Grammar School, etc. There is considerable business done in agricultural produce for the Waterford, Bristol, Limerick, &c., Markets. Fairs, first Wednesday of every month, May 5 and November 5. Markets, Tuesday and Saturday. The borough sends one member to the House of Commons; and it gives the title of Earl to the Scott family. Corporation revenue, £822. It is also the head of the Poor Law Union. Is also the name of a parish of County Cork, Munster, 10 miles East Southeast of Cork. Area, 3,197 acres. Population, 2,564. It comprises part of the town of Cove.

CLONMELLON.—Is a parish in the Western part of County Meath, Leinster, 5 miles Northwest of Athboy. Population, 859. There are three Fairs held here Annually.

CLONMINES.—Is a parish of County Wexford, Leinster, on Bannon Harbor, 7 miles South Southwest of Tagmon. Area, 1,380 acres. Population, 377. The remains of the ruined town of Clonmines are here.

CLONMORE.—There are several parishes by this name, as follows, viz: CLONMORE (or KILLAVENSCH) is in County Tipperary, Munster, 4 miles Northeast of Templemore. Area, 8,160 acres. Population, 3,557; County Carlow, Leinster, 3 miles South Southeast of Hacketstown. Area, 6,029 acres. Population, 2,335. It gives the title of Baron to the Howard family: County Wexford, Leinster, 4 miles South Southwest of Enniscorthy. Area, 6,767 acres. Population, 4,779. There are picturesque ruins of an old Abbey here; County Kilkenny, Leinster, 5½ miles East Southeast of Carrick-on-Suir. Area, 2,092 acres. Population, 795; County Louth, Leinster, 2½ miles East Northeast of Dunteer. Area, 1,905 acres. Population, 725.

CLONMULSK.—There are two parishes by this name. CLONMULSK (or CLONRUSH) is in County Galway, Connaught, 10½ miles Southwest of Portumna. Area, 11,850 acres. Population, 3,115; CLONMULSK (or CLONMELSH) is in County Carlow, Leinster, 4 miles South of Carlow. Area, 3,147 acres. Population, 675.

CLONOE.—Is a parish of County Tyrone, Ulster, 2½ miles South Southeast of Stewartstown. Area, 12,071 acres. Population, 6,817 Surface is low and marshy. In it are the remains of Mountjoy Castle.

CLONOULTY.—Is a parish of County Tipperary, Munster, 4½ miles Southwest of Holycross. Area, 11,135 acres. Population, 3,855. There are two Fairs held in the village Annually.

CLONPRIEST.—Is a parish of County Cork, Munster, 4 miles Southwest of Youghal. Area, 6,985 acres. Population, 3,658.

CLONTARF.—Is a small town and parish of County Dublin, Leinster, 3 miles East Northeast of Dublin, on the North side of the Bay. Area, 1,190 acres, Population, 2,664 ; of town, 818. The town consists of one pretty good street and has a Church which contains the vaults of the Vernon's, Lords of the Manor, who reside in Clontarf Castle. The battle forming the subject of Gray's ode "The Fatal Sisters" was fought here April 23, 1014, in which the united Danes and Irish were defeated by the troops of Brian Borouch, who was killed in the action.

CLONTIBRET.—Is a parish of County Monaghan, Ulster, 7½ miles North Northwest of Castle-Blayney. Area, 26,564 acres. Population, 16,833. Surface mountainous and boggy.

CLONTURK.—Is a parish of County Dublin, Leinster, 2 miles Northeast of Dublin. Area, 1,244 acres. Population, 2,721. It comprises the town of Ballybough.

CLONTURSKERT.—There are two parishes by this name. CLONTURSKERT (or CLONTHUSKERT) is in County Galway, Connaught, 5 miles North Northwest of Eyrecourt. Area, 15,509 acres. Population, 3,711 ; CLONTURSKERT (CLOONTWISCAR) is in County Roscommon, Connaught, 1½ miles North Northwest of Lanesborough. Area, 7,466 acres. Population, 3,221.

CLOONAFF.—(CLONAFF or CLONCRAFT) is a parish of County Roscommon, Connaught, 5 miles North Northeast of Stokestown. Area, 5,454 acres. Population, 2,853.

CLOONOGHILL (or CLOONACOOL) Is a parish of County Sligo, Connaught, 3½ miles West Southwest of Balleymote. Area, 7,098 acres. Population, 2,588.

CLOYNE.—Is a parish and market town of County Cork, Munster, 4 miles Southwest of Castle-Martyr. Area, 9,969 acres. Population, 6,726 ; of town, 2,200. It was well built but has decayed. There are several old Churches and a School endowed in 1719. Market, Thursday. There are four Fairs held here Annually. There are some valuable Marble Quarries in the vicinity.

COLERAINE.—Is a parliamentary and municipal borough, seaport town and parish of County Londonderry, Ulster, on the river Bann, 4 miles from its mouth and 47 miles North Northwest of Belfast. Area, 4,846 acres. Population, 5,857 ; of town, 5,763. Area of parliamentary borough, 963 acres. It is well built and contains many fine buildings, etc. The harbor has been so much improved that large vessels unload at the dock close by the bridge. They have constant communication by steamer with Liverpool, Glasgow and Fleetwood. Markets on Monday, Wednesday and Friday. There are seven Fairs held here Annually. This borough sends one member to the House of Commons and gives the title of Baron to the Hanger family ; about 1 mile South are traces of a Danish Fort. There are Paper Mills, Tanneries, Bleach Grounds also Salmon and Eel fisheries.

COLLON.—Is a market town and parish of County Louth, Leinster, on a branch of the Boyne, 5½ miles Northwest of Drogheda. Area, 8,813 acres. Population, 3,275 ; of town, 936. It is neatly built and has a good Market House, Parish Church, Cotton Factory and Bleaching Grounds. Stocking and Linen Weaving employ many of the inhabitants. There are two Fairs held here Annually. Markets on Tuesday. Collon House, the seat of the Viscount Ferrard, is here.

COLLOONEY.—Is a market town of County Sligo, Connaught, 5½ miles Southwest of Sligo, on the Owenbeg. Population, 651.

COLLUMBKILL.—There are two parishes of Leinster by this name. One is in County Longford, 3 miles West of Granard. Area, 20,314 acres. Population, 9,273. Another is in County Kilkenny. Area, 4,473 acres. Population, 1,116. , It comprises part of the town of Thomastown.

COMADERRY.—Is a mountain in County Wicklow, Leinster, 3 miles West of Glendalough. Elevation, 2,268 feet.

COMBER.—(or CUMBER) Is a market town and parish of County Down, Ulster, on the West side of Lough Strangford, 8 miles East Southeast of Belfast. Area of parish, 17,419 acres. Population, 9,022 ; of town, 1,964. It is well built and contains some ruins. There are four Fairs held here Annually. It has a trade in Linens and Spirits.

CONG.—Is a small town and parish of County Mayo, Connaught, 9 miles West Northwest of Headford. Area, 37,730 acres ; which includes Lough Corrib. Population, 8,835 ; of town, 364. It has a good Church and curious **remains of an Abbey of the 7th Century.**

CONNAUGHT.—Is a province; being the most Westerly and smallest of the four; having the Atlantic Ocean on its North and West; the provinces of Ulster and Leinster on the East, and the province of Munster on the Southeast. Its greatest length is 86 miles from North to South, and breadth 81 miles. Area, 4,392,000 acres; of which about 2,000,000 acres are waste or water. Population in 1851, 1,418,859. The Western part is divided into numerous peninsulars, the largest of which is Connemara and has numerous islands, as Achill, Innisbegil, Clara, etc. The numerous bays afford commodious harbors. The surface of the Western part is mountainous and has some highly picturesque scenery. The surface of the North and South is mountainous, while in the centre it is a level plain. The Bonnet, Uncon, Arrow, Moy, Clare and Shannon, which forms the Eastern boundary, are the principal rivers. Coal is found in Lough Allen district. The province is divided into five Counties, Mayo and Galway on the West and Sligo, Leitrim and Roscommon on the East. Connaught was formerly a Kingdom of the Irish Heptarchy and ruled by the O'Conners; and in 1590 was divided into Counties and came under English administration. The chief towns of the province are Galway, Roscommon, Sligo, Carrick, Castlebar, Tuam, Ballinasloe and Athlone.

CONVOY.—Is a parish of County Donegal, Ulster, 2½ miles Southwest of Raphoe. Area, 20,082 acres. Population, 5,479. There are two Fairs held here Annually.

CONWALL.—(or CONEWAL) Is a parish of County Donegal, Ulster. Area, 45,270 acres. Population, 12,666. It comprises the town of Letter Kenny.

COOTEHILL.—Is a market town of County Cavan, Ulster, on the Cootehill river, 28 miles West Northwest of Dundalk. Population, 2,425. It has some good buildings and a brisk trade in Linens, Corn, Beer and Spirits. Markets, Friday and Saturday, and cattle markets monthly. There are four Fairs held here Annually. Quarter Sessions at Easter and in October.

COOKSTOWN.—Is an inland town of County Tyrone, Ulster, on the Ballinderry, 5 miles West Northwest of Stewartstown. It is well built and has some fine buildings, Churches, etc. Markets, Tuesday and Saturday. There are ten Fairs held here Annually.

There is a parish of same name in County Meath, Leinster, 2 miles East Northeast of Ratoath. Area, 1,238 acres. Population, 142.

COPELAND ISLANDS.—Are a small group of islands off the coast of County Down, Ulster, on the South side of the entrance to Belfast Lough. On Cross island, about 5 miles North Northeast of Donaghada, is a Light-house at an elevation of 131 feet. Latitude, 54°, 4', 44'' North. Longitude, 5°, 32' West.

CORBALLY.—There are several parishes by this name, as follows, viz : Partly in County Kings, Leinster, and partly in County Tipperary, Munster, near Roscrea. Area, 12,747 acres. Population, 3,373 ; County Waterford, Munster, 6¼ miles South Southeast of Waterford. Area, 725 acres. Population, 315 ; County Cork, Munster, 5 miles Southwest of Cork. Area, 869 acres. Population, 193.

CORK.—Is a County of Munster. It is the most Southerly and largest County of Ireland, having County Limerick on the North, County Tipperary on the Northeast, County Waterford on the East, and the Atlantic Ocean on the South and West. Its greatest length is 100 miles, breadth 55 miles. It has an Area of, 1,846,333 acres ; of which 465,889 is waste. Population in 1851, 648,903. In the Western part the surface is very mountainous but in the North and East it is rich and fertile. On the coast are some of the finest bays and harbors in the world, the principal ones being Bantry and Dunmanus bays, and Clonakilty, Kinsale, Cork and Youghal harbors. The Blackwater, Lee and Bannon are the principal rivers. The principal productions of the country are Oats, Wheat and Potatoes ; considerable quantities of which are exported. Average rent of land, 13s. 7d. per acre. Linens, Weaving and Distilling are the principal manufactures. The County is divided into East and West Ridings, 19 baronies and 269 parishes. The County sends two members to the House of Commons. Large quantities of Copper is found ; after which, Limestone is the principal mineral product.

CORK.—Is a city, parliamentary borough and river port of County Cork, Munster, on the river Lee, 137 miles Southwest of Dublin. Area of city, 48,006 acres ; of the municipality, 2,683 acres. Population of city, 106,055 ; of municipal borough, 84,114, most of whom are Roman Catholics. The city proper is built on an island formed by the river Lee, which is here crossed by nine bridges, several of them elegant structures. Its main streets are

50

broad, well paved and lighted with gas, but the largest part of the city consists of lanes inhabited by the very lowest classes. There are many handsome and costly buildings, among which the City and County Court Houses may be said to rank first, having cost £22,000 to erect. There are many other fine buildings, both public and private, also Scientific Institutions, among which are Queens College, Cork Library, etc. There are three Newspapers published here. Its environs are studded with country residences belonging to merchants and others. The principal manufactures are of Leather, Iron and other metallic goods, also Glass, Gloves and Paper; there are also extensive Breweries and Distilleries. It has communication with London, Dublin, Bristol, Liverpool and Glasgow, and the Steamers between New York and Liverpool touch here. It is the seat of Assizes for the City and County of Cork, of Quarter Sessions and Recorder's Weekly Court, and is the headquarters for the Southern military district of Ireland. It sends two members to the House of Commons. Markets daily. Cork is the birthplace of the artists Barry and Maclise, and of the dramatic author, Sheridan Knowles.

CORK HARBOR.—Is a fine land locked harbor which is navigable to 1½ miles above the city of Cork. It is very large and has depth of water for any vessel. Latitude, 51°, 50′. 4″ North. Longitude, 8°, 19′ West. On its shore are the towns of Cove and Passage, with Docks 4 miles in length, which cost £100,000.

CORRIB.—(LOUGH) Is one of the largest lakes in Ireland. It is in County Galway, Connaught, 3 miles North of Galway. Length, 20 miles Northwest to Southwest. Breadth, from 1 to 6 miles. Area, 43,485 acres. The towns of Cong and Oughterard are on its shores. It receives the Clare river and surplus water of Loughs, Mark and Carra, and discharges its own surplus into Galway bay, by the Galway river.

COVE OF CORK.—(now QUEENSTOWN) Is a seaport and market town of County Cork, Munster, 10 miles East Southeast of Cork, on the South side of Cove island in Cork harbor. It has some fine buildings, and a Dock and Station House for pilots and the officers of the port. There is a Light-house on Roche's point at the entrance to Cork harbor. Latitude, 51°, 47′, 35″ North. Longitude, 8°, 13′, 14″ West. Cove is protected by formidable batteries, and opposite it are several islets with additional fortifications. It is much resorted to in summer for bathing.

A SOUTH-EAST VIEW OF THE ROCK OF CASHEL.

COVE, (ISLE OF).—Area, 13,000 acres. Is fertile and connected to the mainland by several bridges. Is also the name of a maritime village of County Cork, Munster, 1 mile East of Kinsale. Population, 352.

CREAGH.—There are two parishes by this name. One is in County Cork, Munster. Area, 5,802 acres. Population, 6,415. It comprises a part of the town of Skibbereen; another is in County Roscommon, Connaught. Area, 8,868 acres. Population, 2,888. It comprises a part of the town of Ballinasloe.

CREGGAN.—Is a parish of Armagh and Louth, Ulster, 9 miles Northwest of Dundalk. Area, 24,815 acres. Population, 15,502.

CROAGH.—Is a parish of County Limerick, Munster, 3½ miles North Northeast of Rathkeale. Area, 7,221 acres. Population, 3,185; of village, 187. Its church was formerly collegiate. There are four Fairs held here Annually.

CROAGH PATRICK.—(or REEK) Is a mountain of County Mayo, Connaught, 6 miles West Southwest of Westport, on the South side of Clew Bay. Elevation, 2,530 feet. It is an object of superstition by the Irish.

CHROGAN KINSHELA.—Is a mountain range of County Wicklow, Leinster. Summit 2,064 feet above the sea. It is famous as the site of the Wicklow Gold Mines.

CROGHAN.—Is a parish of County Kings, Leinster. Population, 915.

CROOM.—Is a town and parish of County Limerick, Munster, 5 miles South Southeast of Adare. Area of parish, 13,437 acres. Population, 7,097; of town, 1,470. There is a strong Castle here, built during the reign of King John. There are four Fairs held here Annually.

CROSSBOYNE.—Is a parish of County Mayo, Connaught, 2 miles South of Claremorris. Area, 16,234 acres. Population, 6,702.

CROSSMAGLEN.—Is a neat market town of County Armagh, Ulster, 10 miles Northwest of Dundalk. Population, 546. Market, weekly. There are six Fairs held here Annually.

CROSSMOLINA.—Is a market town and parish of County Mayo, Connaught, on the river Deel, 6½ miles Southwest of Ballina. Area of parish, 67,201. Surface, mountainous. Population, 12,221; of town, 1,672. The town is well built and contains some good buildings, and remains of an old Castle and Abbey of the Tenth Century.

CRUIT ISLAND.—Is an island in the Atlantic Ocean, off the Northwest coast of County Donegal, Ulster, 6 miles North Northwest of Dungloe. Length, 2 miles from North to South.

CRUMLIN.—Is a market town of County Antrim, Ulster, 12 miles West Northwest of Belfast. Population, 568. Markets, first Friday in every month. There are two Fairs held here Annually. Glendarragh House, the seat of Colonel Heyland, lord of the manor, is near the town. Is also the name of a parish of County Dublin, Leinster, 2½ miles Southwest of Dublin. Area, 1,807 acres. Population, 1,024.

CULDAFF.—Is a parish of County Donegal, Ulster, 9 miles North Northwest of Moville. Area, 20,089 acres. Population, 5,883. The remains of an ancient fort are to be seen here.

CULLEN.—There are several parishes by this name, as follows, viz: County Cork, Munster, 6 miles Northwest of Mill Street. Area, 13,674 acres. Population, 5,490. The remains of several Baronial Castles are here; County Cork, Munster, 5 miles North Northeast of Kinsale. Area, 4,250 acres. Population, 1,330; County Tipperary, Munster, 5 miles Northwest of Tipperary. Area, 1,986 acres. Population, 1,013; of village, 275.

CULLENS WOOD.—Is a suburb of Dublin, on the Southeast side of County Dublin, Leinster. Population, 546. There are some hansome residences here.

CURRIN.—Is a parish of Counties, Monaghan and Fermanagh, Ulster, 5 miles South Southeast of Clones. Population, 6,928.

CUSHENDALL.—(or NEWTOWNGLENS) Is a small market town of County Antrim, Ulster, on the river Dall, 32 miles North of Belfast. Population, 545. It is much frequented as a watering place.

DALKEYS.—Is a maritime parish of County Dublin, Leinster, 8 miles Southeast of Dublin. Area, 467 acres. Population, 1,449; of village, 304. The village is on the Irish sea, outside of Dublin Bay. Its harbor is protected by seven strong forts, now partly dismantled. Dalkey Island is seperated from the mainland by a narrow sound.

DENN.—Is a parish of County Cavan, Ulster, 4 miles Southeast of Cavan. Area, 11,600 acres. Population, 6,696.

DERG LOUGH.—Is the largest and most picturesque of the

Loughs of the Shannon, and separates County Tipperary, Munster, from County Galway, Connaught. Length, 24 miles North to South. Breadth, from 2 to 6 miles. Area, 29,570 acres. It receives the surplus waters of Loughs, O'Grady and Craney Is also the name of a Lough of County Donegal, Ulster, at its Southeast extremity, about 9 miles in circumference, and contains the famous inlet termed St. Patrick's Purgatory, which is annually visited by 18,000 devotees.

DERRALOSSORY.—Is a parish of County Wicklow, Leinster, 8 miles North of Rathdrum. Area, 45,966 acres. Population, 4,897.

DERRY.—Is a parish and city of County Londonderry, Ulster, and the prefix of several parishes. DERRYAGHY is a parish of County Antrim, Ulster, 2 miles North of Lisburn. Area, 12,480 acres. Population, 5,397; DERRYKEIGHAN is in County Antrim, Ulster, 5 miles North Northeast of Ballymoney. Area, 7,643 acres. Population, 3,157; DERRYLORAU is in County Tyrone, Ulster. Area, 12,100 acres. Population, 8,480. It comprises part of the town of Cookstown; DERRYNOOSE (or MADDAN) is in County Armagh, Ulster, 4 miles South Southwest of Keady. Area, 15,049 acres. Population, 9,089; DERRYOULLEN is in County Fermanagh, Ulster. Area, 23,646 acres, including loughs. Population, 10,675. It comprises the town of Lowtherstown.

DERROCK.—Is a neat village of County Antrim, Ulster, 4 miles North Northeast of Ballymoney. Population, 545.

DESERT.—Is a prefix of several parishes. DESERTCREIGHT is in County Tyrone, Ulster, 1½ miles Southwest of Cookstown. Area, 14,399 acres. Population, 7,675. Linen weaving is the principal manufacture. DESERTLYN is in County Londonderry, Ulster. Area, 5,561 acres. Population, 3,255. It comprises part of the town of Moneymore; DESERTMARTIN is in County Londonderry, Ulster, 3 miles Southeast of Tubbermore. Area, 9,580 acres. Population, 5,023; of village, 256; DESERTOGHILL is in County Londonderry, Ulster, 4 miles West Northwest of Kilrea. Area, 11,469 acres. Population, 4,901; DESERTSERGES is in County Cork, Munster, 6 miles West Southwest of Bandon. Area, 15,730 acres. Population, 6,327.

DEVENISH.—Is a parish of County Fermanagh, Ulster, 5 miles North Northwest of Enniskillen. Area, 32,243 acres, including Loughs. Population, 8,381

DEVILS BIT MOUNTAINS.—Are a range of mountains in County Tipperary, Munster, extending from Northeast to Southwest, a length of 24 miles, and separating the basins of the Shannon and Suir rivers. Height, 2,084 feet.

DONABATE.—Is a village and parish of County Dublin, Leinster, 10½ miles North Northeast of Dublin. Area, 2,715 acres. Population, 479. It is a station on the Dublin and Drogheda railroad.

DONAGH.—There are two parishes by this name. One is in County Monaghan, Ulster, 5 miles North Northeast of Monaghan. Area, 16,202 acres. Population, 10,244; another is in County Donegal, Ulster. Area, 25,259 acres. Population, 5,447. It comprises the town of Earn.

DONAGHADEE.—Is a seaport, market town and parish of County Down, Ulster, on the Irish channel, 16¼ miles East Northeast of Belfast. Area, 9,593 acres. Population, 8,557; of town, 3,151. It is well built and has a good harbor, also a light-house. There are quite a number of Flax Mills in the town. Communication to Portpatrick by Steamer. There are five Fairs held here Annually.

DONAGHCLONEY.—Is a parish of County Down, Ulster, 2½ miles Southeast of Lurgan. Area, 6,698 acres. Population, 6,373.

DONAGHEADY.—Is a parish of County Tyrone, Ulster, 7½ miles Northeast of Strabane. Area, 39,398 acres. Population, 10,608.

DONAGHEAVY.—(or FINDONACH) Is a parish of County Tyrone, Ulster. Area, 23,052 acres. Population, 11,229. It comprises the town of Fintona. There is a great deal of Linen weaving done here.

DONAGH-HENRY.—Is a parish of County Tyrone, Ulster. Area, 7,155 acres. Population, 5,673. It includes Stewartstown and part of Coal Island.

DONAGHMORE.—There are numerous parishes by this name, as follows, viz: County Donegal, Ulster, 4 miles Southwest of Lifford. Area, 46,378 acres. Population, 12,955, (including the town of Castle-Linn); County Tyrone, Ulster, 2½ miles Northwest of Dungannon. Area, 18,410 acres. Population, 12,333; of village, 542. There are twelve Fairs held here Annually; County Cork, Munster, 11½ miles East Northeast of Macroom. Area, 22,308 acres. Population, 7,491. It gives the title of Earl to the Hely

Hutchison family ; County Down, Ulster, 4½ miles North of Newry Area, 8,396 acres. Population, 4,436 ; County Wicklow, Leinster, 4 miles North Northeast of Battinglass. Area, 23,428 acres. Population, 3,910 ; County Wexford, Leinster, 6 miles South Southeast of Gorey, Area, 7,478 acres. Population, 2,497 ; County Queens, Leinster, 2½ miles North Northwest of Rathdowney. Area, 3,529 acres. Population, 1,620 ; of village, 496 ; County Meath, Leinster. Area, 3,955 acres. Population, 1,524. It comprises part of the town of Navan. ; County Limerick, Munster, 2½ miles South Southeast of Limerick. Area, 943 acres. Population, 727 ; County Cork, Munster, 7½ miles East Southeast of Clona. Area, 312 acres. Population, 458 ; County Meath, Leinster, 4 miles East Southeast of Ratoath. Area, 3,413 acres. Population, 391 ; County Kildare, Leinster, 1½ miles East Southeast of Maymooth. Area, 301 acres. Population, 29.

DONAGHMOYNE.—Is a parish of County Monaghan, Ulster, 3 miles North Northeast of Carricknacross. Area, 25,604 acres. Population, 15,100.

DONAGHPATRICK.—There are two parishes by this name. One is in County Galway, Connaught, 3 miles East Northeast of Headford. Area, 10,342 acres. Population, 3,770 ; another is in County Meath, Leinster, 4½ miles Northwest of Navan. Area, 4,028 acres. Population, 786.

DONEGAL. Is a maritime County of Ulster. The Counties, Tyrone, Londonderry and Fermanagh are on the South and East, and on the North and West the Atlantic Ocean and Donegal Bay. Area, 1,185,641 acres ; of which 769,587 are waste. Population in 1851, 251,288. The surface is mountainous. The Swilly and Leenan are the principal rivers, and Swilly and Mulroy the principal Loughs. The soil is not very fertile and the chief productions are Oats, Barley, Flax and Potatoes. Average rental, 13s. 7d. The fisheries are very extensive, and employ 13,700 hands and 3,000 vessels. Linen is the principal manufacture. The County is divided into 6 baronies and 51 parishes. Ballyshannon and Letterkenny, with the ports of Rathmelton, Donegal and Killybegs are the principal towns. It sends two members to the House of Commons.

DONEGAL BAY.—Is an inlet of the Atlantic Ocean, extending inland about 25 miles, with a breadth at its entrance of 20 miles.

DONEGAL.—Is a seaport, market town and parish of County

Donegal, Ulster, at the mouth of the Esk, in Donegal Bay, 11 miles North Northeast of Ballyshannon. Area of parish, 23,260. Population, 6,588; of town, 1,366. Vessels drawing 12 feet enter its harbor. There are six Fairs held here Annually.

DONERAILE.—Is a market town and parish of County Cork, Munster, on the Awbeg river, 6 miles North Northeast of Mallow. Area of parish, 20,442 acres. Population, 8,350; of town, 2,722. It is a poor place with some public buildings. Market on Saturday. Fairs, 12th of August and 12th of November. It gives the title of Viscount to the St. Leger family.

DONISLE.—(or DUNHILL) Is a parish of County Waterford, Munster, 6 miles Southeast of Kilmacthomas. Area, 6,287 acres. Population, 2,160. Here are the ruins of the Castle Don-Isle destroyed by Cromwell.

DONNYBROOK.—(ST. MARY'S) Is a parish of County Dublin, Leinster, 2 miles Southeast of Dublin. Area, 1,689 acres. Population, 9,825; of village, 1,610. Including the small towns of Irishtown, Ringsend and Sandymount. The latter on the Dodder has some fine buildings and several mills, but is chiefly famous for its Fair held during the week commencing August 26.

DONOHILL.—Is a parish of County Tipperary, Munster, 4 miles North of Tipperary. Area, 13,914 acres. Population, 4,834.

DOOISH.—Is a mountain in County Donegal, Ulster, 10 miles West Northwest of Letterkenny. Height, 2,143 feet above the sea.

DOON.—Is a parish of County Tipperary, Munster, 9 miles North Northwest of Tipperary. Area, 27,630 acres. Population, 7,895.

DOUCE.—(vulg DJOUCE) Is a mountain in County Wicklow, Leinster, 5½ miles South Southwest of Bray. Elevation, 2,392 feet.

DOWN.—Is a maritime County of Ulster, having Counties, Antrim and Armagh on the West and on other sides the Irish Sea and Belfast Lough. Area, 611,209 acres; of which about 514,000 are arable. Population in 1851, 325,575. It is separated on the South by Carlingford Bay, and on the West by the Newry Canal and Lagan river. The Bann and Anahill rivers are also in this County. Lough Strangford, Dundrum Bay and the Mourne mountains are also in this County. Surface is generally mountainous, but in some places it is very fertile. The rural population are said to be better off in this than in many other Irish Counties. The principal crops are Potatoes, Oats, Barley and Flax. The fisheries are very extensive, employing about 14,000 hands. Linens, Muslins, Leather, Hosiery, Cotton Thread and Salt are the chief manufactured products, which are exported with Butter, Corn, Pork, Hide and Skins. This County is subdivided into 10 baronies and 70 parishes in the diocese of Down and Dromore. Downpatrick, Newton-Ardes and Newry are the principal towns. This County returns two members to the House of Commons.

DOWNPATRICK.—Is a municipal and parliamentary borough, seaport town and parish of County Down, Ulster, near the mouth of the Quoyle, in Lough Strangford, 21 miles South Southeast of Belfast. Area of parish, 11,485 acres. Population, 8,812. Area of borough, 1,487 acres. Population, 4,866 ; of town, 4,651. It is well built and consists chiefly of four streets, and is divided into English, Irish and Scotch quarters, and contains some fine public buildings. Market, Saturdays. There are twelve Fairs held here Annually. Near the town are the ruins of the old Cathedral, a remarkable ancient mound, a race course on which races take place every other July, and was much resorted to by Roman Catholic Pilgrims. It is the seat of County Assizes, Quarter and Petty Sessions, and sends one member to the House of Commons.

DRIMOLEAGUE.—(or DROMDALEAGUE) Is a parish of County Cork, Munster, 8 miles East Southeast of Bantry. Area, 18,708 acres. Population, 5,501.

DRIMTEMPLE.—(or DRUMOTEMPLE) Is a parish of Counties, Galway and Roscommon, Connaught, 4 miles South Southeast of Castlerea. Area, 6,531 acres. Population, 2,878.

DRINAH.—There are two parishes by this name. One is in County Cork, Munster, 3½ miles South of Dunmanway. Area, 12,869 acres. Population, 4,598. Another is in County Wexford, Leinster, 1½ miles South Southeast of Wexford. Area, 1,171 acres. Population, 436.

DRISHANE.—Is a parish of County Cork, Munster. Area, 33,085 acres. Population, 8,868. It comprises the town of Millstreet.

DROGHEDA.—Is a parliamentary and municipal borough, seaport and town of Counties, Meath and Louth, Leinster, on both sides of the Boyne river, 4 miles from its mouth, and 31½ miles North of Dublin. Area, 497 acres. Population in 1851, 16,845 ;

of parliamentary borough, 17,300. It is generally well built and contains many fine buildings and does considerable manufacturing. The harbor and river is much improved, and vessels of 200 tons can ascend the river to the bridge. Markets, Thursdays and Saturdays. There are eight Fairs held here Annually. Assizes, Quarter and Petty Sessions are held here, and the borough sends one member to the House of Commons. It gives the title of Marquis to the Moore family.

DROMAGH.—Is a village of County Cork, Munster, 5½ miles Southwest of Kanturk. There are large Collieries, Bolting Mills, and a great quantity of coarse Pottery is manufactured.

DROMAHAIRE.—Is a barony and village of County Leitrim, Connaught, 7½ miles East Southeast of Sligo. Population, 348.

DROMARAGH.—Is a parish of County Down, Ulster, 6 miles East Southeast of Dromore. Area, 21,192 acres. Population, 10,070; of village, 226.

DROMCLIFFE.—(or Ogoomtck) Is a parish of County Clare, Munster. Area, 9,968 acres. Population, 13,211. It comprises the town of Ennis.

DROMIN.—There are two parishes by this name. One is in County Limerick, Munster, 3 miles South Southwest of Bruff. Area, 4,096 acres. Population, 1,375. Another is in County Louth, Leinster, 1½ miles West Northwest of Dunleer. Area, 2,042 acres. Population, 863.

DROMID.—Is a parish of County Kerry, Munster, 7 miles South Southeast of Cahirciveen. Area, 50,702 acres. Population, 5,247. The surface is rugged.

DROMISKIN.—Is a parish of County Louth, Leinster, 2 miles North of Castle-Bellingham. Area, 5,312 acres. Population, 2,636.

DROMORE.—Is a parish and city of County Down, Ulster, on the Lagan river, 14¾ miles Southwest of Belfast, on the road to Dublin. Area, 20,488 acres. Population, 14,954; of city, 2,110. The town is well built and has many good buildings, and a good trade in Linens, as many of the inhabitants are engaged in its manufacture. It has County Petty Sessions. There are eight Fairs held here Annually. The Diocese comprises 27 parishes, in Counties, Down, Antrim and Armagh. Is also the name of a parish of County Tyrone, 8½ miles Southwest of Omagh. Area, 25,492 acres. Population, 10,601.

DROUMTARIFF.—Is a parish of County Cork, Munster, 4

miles South Southwest of Kanturk. Area, 15,224 acres. Population, 7,261.

DRUM.—There are two parishes by this name. One (or ELARDRIUM) is in County Roscommon, Connaught, 5½ miles Southwest of Athlone. Area, 16,149 acres. Population, 5,048. Another is in County Mayo, Connaught, 5 miles Southeast of Castlebar. Area, 7,768 acres. Population, 4,127.

DRUMACHOSE.—Is a parish of County Londonderry, Ulster. Area, 11,685 acres. Population, 5,463. It comprises the town of Newton-Limavaddy.

DRUMAUL.—Is a parish of County Antrim, Ulster. Area, 32,394 acres. Population, 9,818. It includes the town of Randalstown.

DRUMBALLYRONEY.—Is a parish of County Down, Ulster. Area, 12,339 acres. Population, 9,138. It comprises part of the town of Rathfriland.

DRUMBOE.—Is a parish of County Down, Ulster, 4 miles East of Lisburn. Area, 13,793 acres. Population, 8,271.

DRUMCANNON.—Is a parish of County Waterford, Munster, 7 miles South of Waterford. Area, 7,672 acres. Population, 3,988. It includes the town of Tramore.

DRUMCLIFFE.—Is a parish of County Sligo, Connaught, 4 miles North of Sligo. Area, 26,590 acres. Population, 12,982.

DRUMCOLLIKER.—Is a parish of County Limerick, Munster, 10¼ miles West of Charleville. Area, 4,846 acres. Population, 2,781.

DRUMCONDRA.—(or DRUMCONRA) Is a parish of County Meath, Leinster, 4 miles East Northeast of Nobber. Area, 7,926 acres. Population, 3,013 ; of village, 419. There are four Fairs held here Annually. Is also the name of a village of County Dublin, Leinster, 2 miles North of Dublin. Population, 227.

DRUMCOREE.—Is a parish of County Armagh, Ulster. Area, 13,386 acres. Population, 14,038. It comprises the town of Portadown. Is also the name of a village of County Meath, Leinster.

DRUMCULLEN.—Is a parish of County Kings, Leinster, 5 miles South Southwest of Ballyboy. Area, 13,904 acres. Population, 4,608.

DRUMGATH.—Is a parish of County Down, Ulster. Area, 5,331 acres. Population, 4,608, with town of Rathfriland.

DRUMGLASS.—Is a parish of County Tyrone, Ulster. Area, 3,504 acres. Population, 6,089, with town of Dungannon.

DRUMGOOLAND.—Is a parish of County Down, Ulster. Area, 19,653 acres. Population, 10,567.

DRUMGOON.—Is a parish of County Cavan, Ulster. Area, 15,475 acres. Population, 12,575, with town of Cootehill.

DRUMHOLM.—(or Drumhome) Is a parish of County Donegal, Ulster, 4 miles North of Ballyshannon. Area, 35,433 acres. Population, 9,893.

DRUMKEERAN.—Is a parish of County Fermanagh, Ulster, 1½ miles North Northwest of Kesh. Area, 27,159 acres. Population, 8,751. It comprises part of the town of Pettigoe. Is also the name of a village of County Leitrim, Connaught, 5½ miles South Southeast of Dromahaire. Population, 469.

DRUMLANE.—Is a parish of County Cavan, Ulster. Area, 20,066 acres. Population, 9,438.

DRUMLEASE.—Is a parish of County Leitrim, Connaught, 6½ miles Northwest of Sligo. Area, 15,271 acres. Population, 8,807.

DRUMLOMAN.—Is a parish of County Cavan, Ulster, 2½ miles East Northeast of Granard. Area, 17,248 acres. Population 8,807.

DRUMMULLY.—Is a parish of Counties, Monaghan and Fermanagh, Ulster, 4 miles West Southwest of Clone. Area, 7,547 acres. Population, 4,534.

DRUMQUIN.—Is a village of County Tyrone, Ulster, 7½ miles West of Omagh. Population, 452. There are eight Fairs held here Annually.

DRUMRAGH.—Is a parish of County Tyrone, Ulster. Area, 20,164 acres. Population, 11,453. It comprises the town of Omagh.

DRUMRANEY.—(or Drumrath) Is a parish of County Westmeath, Leinster, 8½ miles Northeast of Athlone. Area, 9,102 acres. Population, 3,367.

DRUMREILLY.—Is a parish of County Leitrim, chiefly in Connaught, 4 miles East Northeast of Ballinamore. Area, 33,673 acres. Population, 10,289.

DRUMSHAMBO.—Is a village of County Leitrim, Connaught, 4½ miles North of Leitrim, on the Shannon river. Population, 517.

DRUMSNA.—Is a small but stirring market town of County Leitrim, Connaught, on the Shannon river, 4½ miles East Southeast of Carrick. Population, 516.

DRUMSNAT.—Is a parish of County Monaghan, Ulster, 4½ miles South Southwest of Monaghan. Area, 5,019 acres. Population, 3,439.

DRUNG.—Is a parish of County Cavan, Ulster, 6 miles East Northeast of Cavan. Area, 1,475 acres. Population, 6,551.

DUBLIN.—Is the capital city of Ireland, County Dublin, Leinster. It is situated on the Liffey river, near its entrance into Dublin Bay, 66 miles West of Holyhead and 135 miles West of Liverpool. Latitude of its observatory, 53° 23' 2" North. Longitude, 6° 20' 5" West. Population in 1851, 287,729. The city proper is nearly surrounded by the circular road, 9 miles in length, and which (with a branch of the Grand Canal on the South and Southeast) encloses an area of 1,264 acres, intersected from West to East by the Liffey river, which is here crossed by a number of bridges and bordered by Granite docks 2½ miles in length.

DUBLIN CASTLE.—An edifice of different ages, built on a hill and containing an Arsenal, an Armory, the Vice-regal Chapel and various government offices. The state apartments of the Lord Lieutenant of Ireland, who, however, resides generally at a mansion in Phœnix Park, is almost in the centre of the Southern half of the city. Adjoining the castle on the South side are its gardens, and on its North side is the elegant Exchange. A line of streets extending from which may be considered, with the Liffey, divide Dublin into four parts. The Southwestern part or quarter is mostly ill-built, irregular and filthy; in this part are Christ Church and St. Patrick's Cathedrals, the Archbishop's Palace, and the Royal and Foundling Hospitals. The Northwestern quarter is much better built than the former, and is principally occupied by the trading and middle classes. On its outskirts are many good edifices, including the new House of Industry, Richmond Penitentiary and the new Courts of Law erected on King's Inn Quay. The Eastern quarter is the handsomest; and the finest approach is from the Northeast. Sackville Street, one of the finest thoroughfares in Europe, the Post Office and Rotunda, are in this portion, and in its centre, the Nelson Pillar, a Doric Column, 134 feet in height, and surmounted by a Statue. The Custom House and Marlboro' Green are also in this quarter. The Southeast quarter comprises St. Stephen's Green, in which is the equestrian statue of George II., Merrion and Fitzwilliam Squares, College Green, (now a paved area, on which is the Bank of Ireland,) Trinity College, a Bronze Statue of William III., Dame street, Sir P. Dun's Hospital, the Dublin Royal Society House and the Mansion House, near which is an equestrian Statue of George I. There are many other handsome buildings, among which are St. George's Church, which has a steeple 200 feet in height. There are 29 Protestant, 9 Roman Catholic and many other Churches of the different denominations, the Commercial Buildings, Corn Exchange, Royal Hospital at Kilmainham, various Barracks, the new Inns of Court, the County Infirmary, Richmond Surgical, Dublin, St. Vincent's, Steven's, Mercer's, and several other general hospitals, the General Dispensary, the Richmond Lunatic Asylum and numerous other Charitable Institutions, the city Assembly House, Corporate Halls, Sessions House, Newgate, and several other prisons, and several Theatres. Its University, chartered in 1591, is situated in Trinity College, and is attended by about 2,000 students, and possesses a landed revenue of £15,000 a year, a Library of 150,000 volumes, a Park, Printing House, Anatomical and Chemical Departments, a Botanic Garden and an Observatory. There are also Colleges of Physicians and Surgeons, an Apothecaries Hall, Royal, Irish and Hibernian Academies, a Royal Institution, Zoological, Geological, Phrenological, Agricultural, Horticultural and other Societies. The Zoological society has gardens in the Phœnix Park, a fine open space at the West extremity of the city, in which is the Wellington Testimonial, on a heavy obelisk, raised at a cost of £20,000. The villages of Ringsend, Irishtown, Sandymount, Ballesbridge, Donnybrook, Ranelagh, Rathmines, Harold's Cross. Kilmainham, Glassnevin, Drumcondra and Clontarf, are suburbs of Dublin. Dublin has communications with all points, by the Grand and Royal Canal, by railroad, and inland and Ocean steamers. It has a fine harbor of 3,030 acres, and near the mouth of the Liffey are the Grand Canal and Custom House docks, the latter occupying 8 acres; depth at low water, 12 feet, at high water, 24 feet; the wharfs and docks are accessible to vessels of 900 tons. There was formerly a large trade done in Woolen, Silk and Cotton Fabrics, but the principal trade now is in the export of Linens, Poplins, Porter, Provisions, &c. The Corporation consists of the Lord Mayor and

15 Aldermen and 45 town Counsellors, elected out of the 15 Municipal Wards. Besides the Lord Mayor's weekly Courts, there are Courts of Conscience for debts under 40 shillings ; and 5 Manor Courts. The See of the Archbishop comprises the Counties, Dublin and Wicklow, with part of Kildare, and his jurisdiction is nearly co-extensive with the provinces of Leinster and Munster. Dublin is divided into 20 parishes and 15 Municipal Wards, and sends two members to the House of Commons. It is also the seat of a Chamber of Commerce and the Ouzel Gallery Society for the arbitration of commercial disputes.

DUBLIN BAY.—Is a bay of County Dublin, Leinster; situated between Howth Head on the North, with Baily Light-house, Latitude 53° 21' 40" North, Longitude, 6° 3' 5" West, and Kingston on the South, with the Light-house. Latitude, 53° 18' North. Longitude, 6° 8' West. The villages of Ratheny, Clontarf, Blackrock and Kingstown are on its shores, which are bold and picturesque.

DUBLIN.—Is a County of Leinster, having the Irish sea on the East, and from North to South having the Counties, Wicklow, Kildare and Meath. Area, 226,414 acres ; of which 19,312 acres are uncultivated, and 1,820 acres are in towns, excluding Dublin city. Population in 1851, 146,631, excluding the capital. The Liffey is the principal river, and the chief towns are Dublin, Kingstown, Blackrock, Balbriggan and Swords. The surface is level and the County is much sub-divided. The fisheries and manufactures are very important. This County sends two members to the House of Commons.

DULEEK.—Is a disfranchised borough, market town and parish of County Meath, Leinster, on the Nanny Water river, 5 miles South Southwest of Drogheda. Area of parish, 16,554 acres. Population, 5,594 ; of town, 1,158. It has the remains of an Abbey of the 12th Century.

DUMANWAY.—Is a market town of County Cork, Munster, near the junction of the three streams which form the Bandon river, 29 miles West Southwest of Cork. Population, 3,086. It has a large Church and some public buildings.

DUNAGHY.—Is a parish of County Antrim, Ulster, 5½ miles North of Ballymena. Area, 13,743 acres. Population, 3,881.

DUNAMANAGH.—Is a village of County Tyrone, Ulster, 8 miles East Northeast of Strabane.

DUNBOE.—Is a parish of County Londonderry, Ulster, 5 miles West Northwest of Coleraine. Area, 14,811 acres. Population, 4,627.

DUNBOYNE.—Is a parish of County Meath, Leinster, 9½ miles Northwest of Dublin. Area, 13,686 acres. Population, 2,347.

DUNCANNON.—Is a maritime village of County Wexford, Leinster, on Waterford Harbor, 2 miles South Southeast of Ballyhack. Population, 521. It gives the title of Viscount to the Earl of Besborough.

DUNDALK.—There are two baronies, a parliamentary and municipal borough, seaport town and parish of County Louth, Leinster, on the South bank of Castletown river, near its mouth in Dundalk Bay, 45 miles Northwest of Dublin. Area of parish, 6,202 acres. Population, 13,204. Area of town and borough, 450 acres. Population in 1851, 9,995. It is principally poor and miserably built, but has some good streets and fine buildings and manufactories. Its fisheries are important. Markets, Monday. Fairs, the third Wednesday in each month. It is the seat of County Assizes, Quarter and Petty Sessions, and sends one member to the House of Commons.

DUNDALK BAY.—Is an inlet of the Irish sea of no importance, between Cooley and Dunmany points. It is the outlet of the Dee, Fane and Castletown rivers, and contains large Oyster Beds.

DUNDRUM BAY.—Is a bay of the Irish sea, County Down, Ulster, 7 miles Southwest of Downpatrick, having St. Johns point on its Northeast side and on the Southwest the Mourne mountains. There are also two villages of this name: one in the inner harbor of this bay, another in County Dublin, Leinster, 4½ miles South of Dublin.

DUNEAN.—Is a parish of County Antrim, Ulster, 8½ miles West Northwest of Antrim. Area, 30,128 acres. Population, 6,369.

DUNFANAGHY.—Is a market town of County Donegal, Ulster, on the South side of Dunfanaghy harbor, 12½ miles Northwest of Kilmacrenan.

DUNFEENEY.—(or DOONFEENY.) Is a parish of County Mayo, Connaught, 10 miles Northwest of Killala. Area, 31,251 acres. Population, 4,819.

DUNGANNON.—Is a parliamentary and municipal borough,

and market town of County Tyrone, Ulster, near a branch of the Blackwater, 11 miles North Northwest of Armagh. Area, 230 acres. Population of town and parliamentary borough, 3,801. It is well built and has a good Church and other buildings, and manufactories of Linen and Earthenware. Markets on Monday and Thursday. Fairs, first Thursday of every month. The borough sends one member to the House of Commons. It gives the title of Viscount to the Hill-Trevor family.

DUNGANSTOWN.—Is a parish of County Wicklow, Leinster, 5 miles South Southwest of Wicklow. Area, 14,287 acres. Population, 3,434.

DUNGARVAN.—Is a parliamentary and municipal borough, and seaport town and parish of County Waterford, Munster, 25 miles West Southwest of Waterford, on the Colligan river, near its mouth in Dungarvan bay. Area of parish, 9,413; of borough, 8,499. Population of parish, 13,321; of borough, 12,382; of town in 1851, 6,417. It is neatly built and much resorted to for sea-bathing There are some fine edifices and public buildings. It has a small trade, as its harbor is not fitted for vessels over 150 tons. Markets daily. There are four Fairs held here Annually. It has County, Quarter and Petty Sessions, and sends one member to the House of Commons. It gives the title of Viscount to the Earl of Cork. Is also the name of a parish of County Kilkenny, Leinster, 3 miles West Southwest of Gowran. Area, 5,881 acres. Population, 1,806.

DUNGIVEN.—Is a market town and parish of County Londonderry, Ulster, 16½ miles East Southeast of Londonderry. Area, 29,328 acres. Population, 5,169; of town, 1,016. The town is well built and has some remains of an old Castle. Market, Saturday. There are eight Fairs held here Annually.

DUNLAVAN.—Is a town and parish of County Wicklow, Leinster, 5 miles Northwest of Dunard. Area, 5,852 acres. Population, 2,594; of town, 990.

DUNLECKNEY.—Is a parish of County Carlow, Leinster. Area, 7,956 acres. Population, 4,743.

DUNLEER.—Is a disfranchised borough and parish of County Louth, Leinster, 11 miles South of Dundalk. Area, 2,379 acres. Population, 1,551; of town, 808.

DUNLUCE.—Is a parish of County Antrim, Ulster. Area,

9,381 acres. Population, 3,381. It comprises part of the town of Bushmills.

DUNMANWAY.—Is a market town of County Cork, Munster, 13 miles West of Bandon. Population, 3,086.

DUNMORE.—Is a town and parish of County Galway, Connaught, 7½ miles North Northeast of Tuam. Area, 34,939 acres. Population, 11,775 ; of town, 917. Is also the name of a parish of County Kilkenny, Leinster, 3½ miles West Northwest of Kilkenny. Area, 2,380 acres. Population, 767. Is also the name of a seaport town of County Waterford, Munster, 8½ miles Southeast of Waterford, on Waterford harbor. Population, 302. It is well built and much frequented as a watering place.

DUNSHAUGHLIN.—Is a market town and parish of County Meath, Leinster, 11 miles South Southeast of Navan. Area, 3,264 acres. Population, 1,581 ; of town, 524.

DUNSINSK.—Is a village of County Dublin, Leinster, 4 miles West Northwest of Dublin Castle.

DURAS.—(or KINVARRA-DURRAS.) Is a parish of County Galway, Connaught, 8 miles Northwest of Gort. Area, 11,290 acres. Population, 6,586. It comprises the town of Kinvarra.

DURROW.—Is a market town and parish of Counties, Kilkenny and Queens, Leinster, 5½ miles South Southwest of Abbeyleix. Area, 6,529 acres. Population, 2,977 ; of town, 1,318. It has large Flour Mills. Is also the name of a parish of County Kings, Leinster, 4 miles North of Tullamore. Area, 9,773 acres. Population, 2,922.

DURRUS.—Is a parish of County Cork, Munster, 3 miles South Southwest of Bantry. Area, 11,138 acres. Population, 4,483.

DURSEY.—Is an Island off the Southwest extremity of Munster, between the estuary of the Kenmore river and Bantry bay. Population, 200.

DYSERT.—(or DYSART.) There are numerous parishes by this name, as follows, viz : County Kilkenny, Leinster, 4½ miles Southeast of Castlecomer. Area, 7,938 acres. Population, 2,369 ; County Clare, Munster, 2 miles South of Carrofin. Area, 7,251 acres. Population, 1,933 ; County Roscommon, Connaught, 6¼ miles South Southeast of Mount Talbot. Area, 6,569 acres. Population, 1,793; County Kerry, Munster, 1½ miles South of Castle Island. Area, 6,070 acres. Population, 1,529 ; County Waterford, Munster, 8 miles East of Clonmel. Area, 5,396 acres. Population,

1,406; County Kerry, Munster, 6½ miles Southwest of Listowel. Area, 6,149 acres. Population, 1,295; County Westmeath, Leinster, 5 miles South Southwest of Mullingar. Area, 7,417 acres. Population, 1,129; County Louth, Leinster, 2 miles East of Dunleer. Area, 1,912 acres. Population, 608; County Limerick, Munster, 3½ miles South Southeast of Adore. Area, 910 acres. Population, 170; GALLEN is in County Queens, Leinster. Area, 10,784 acres. Population, 4,342, including the town of Ballinakilly.

EAGLEISLAND.—Is an island in the Atlantic of County Mayo, Connaught, 4 miles West Southwest of Erris-Head. It has two Light-houses. Latitude, 54° 7′ North, Longitude, 10° 6′ West.

EAGLE MOUNTAIN.—County Down, Ulster, is one of the highest of the Mourne mountains Elevation, 2,084 feet.

EAGLESNEST.—(THE) Is a rock of County Kerry, Munster, between the upper and middle lakes of Killarney, 4 miles Southwest of Killarney. It is an almost perpendicular crag, 1,300 feet in height.

EASKEY.—Is a village and parish of County Roscommon, Connaught, 19½ miles South Southwest of Sligo, on river of same name. Area, 13,285. Population, 6,349. There are two Fairs held here Annually.

EASTERSNOW.—Is a parish of County Roscommon, Connaught, 4 miles East Southeast of Boyle. Area, 6,457 acres. Population, 2.035.

EDENDERRY.—Is a market town of County Kings, Leinster, near the bog of Allen, 32½ miles West of Dublin. Population, 1,850.

EDGEWORTHSTOWN.—Is a parish of County Longford, Leinster, 6¾ miles East Southeast of Longford. Area, 10,943 acres. Population, 4,933. There are six Fairs held here Annually.

EGLISH.—Is a parish of County Kings, Leinster, 3¾ miles North Northeast of Birr. Area, 14,799 acres. Population, 3,494. Is also the name of a parish of County Armagh, Ulster, 4 miles North Northeast of Tynan. Area, 10,500 acres. Population, 5,601. The residence of the Earl of Charlemont "Elm Park" is in this parish. Is also the name of a village of County Tyrone, Ulster.

EIRKE.—Is a parish of Counties, Queens and Kilkenny, Leinster, 4 miles North Northwest of Urlingford. Area, nearly 19,000 acres. Population, 5,678.

ELPHIN. -Is a market town, parish and bishop's See of County Roscommon, Connaught, 17½ miles West Northwest of Longford. Area, of parish, 12,544 acres. Population, 6,781 ; of town, 1,551. There is some of the finest pasturages in the Kingdom in this parish. The diocese comprises 76 parishes in Counties, Roscommon, Galway, Sligo and Mayo, and is now annexed to the Sees of Kilmore and Ardagh. The poet Goldsmith is said by some to have been born here.

ELY.—Is a beautiful demesne of County Fermanagh, Ulster, consisting of several islets at the head of lower Lough Erne, 4 miles North of Enniskillen. It gives the title of Marquis to the Loftus family.

EMATRIS.—Is a parish of County Monaghan, Ulster, 3½ miles West of Ballybay. Area, 12,298 acres. Population, 7,643.

EMLY.—Is a small market town and Episcopal parish of County Tipperary, Munster, 8 miles West Southwest of Tipperary. Area of parish, 9,183 acres. Population, 4,011 ; of town, 650.

EMLYFADD.—Is a parish of County Sligo, Connaught, and 11½ miles South Southwest of Sligo. Area, 9,453 acres. Population, 4,811. It has the ruins of a small Abbey and Castle built in 1300, and comprises the town of Ballymote.

ENFIELD.—Is a village of County Meath, Leinster, 24 miles West Northwest of Dublin.

ENNEL.—(or BELVEDERE.) Is a Lough of County Westmeath, Leinster, 2 miles South Southwest of Mullingar. Area, 3,603 acres.

ENNIS.—Is a parliamentary and municipal borough and market town of County Clare, Munster, on the river Fergus, 20 miles West Northwest of Limerick. Population in 1851, 8,623. It contains all the usual buildings, churches, etc. Markets, Tuesday and Saturday. Fairs, April 25 and September 3. The borough sends one member to the House of Commons.

ENNISCORTHY.—Is a municipal borough and market town of County Wexford, Leinster, 12 miles North Northwest of Wexford. Population about 7,000. It contains a Court House, Bridewell, Roman Catholic Cathedral, etc. Markets, three times a week. Fairs, monthly. General sessions at Easter and Michaelmas. It was the scene of frightful outrages during the rebellion of 1793.

ENNISKENN.—Is a parish of Counties, Cavan and Meath, Ulster and Leinster. Area, 21,000 acres. Population, 11,548.

ENNISKERRY.—Is a village of County Wicklow, Leinster, 11 miles South Southeast of Dublin. Population, 418.

ENNISKILLEN.—Is a parliamentary and municipal borough, market town, parish and capital of County Fermanagh, Ulster, mostly situated on an island in the river connecting upper and lower Lough Erne, 87 miles Northwest of Dublin. Area of parish, 26,500 acres. Population of town, 8,700. It is well built and has a County Court House and Prison, a Town Hall in which is still preserved the banners of the battle of the Boyne, Union Work-House, etc. Markets, Tuesday and Thursday. It sends one member to the House of Commons, and gives the title of Earl to the Cole family, by whom it was founded in 1641. The inhabitants warmly supported the Protestant cause in 1689.

ENNISTRAHUL.—Is a small island off the North coast of County Donegal, Ulster, 7 miles East Northeast of Malin Head. It has a Light-house with a revolving light.

ENNISTYMON.—Is a market town of County Clare, Munster, near the mouth of the river of the same name in Liscanor bay, 14½ miles West Northwest of Ennis. Population, 2,089. It has a Union Work-House, etc. There are seven Fairs held here Annually.

ERNE.—Is a river and two celebrated Loughs of Ireland. The river issues from Lough Gauny, County Cavan, Ulster, and flows into upper and lower Loughs Erne. Length, including the two Loughs, about 60 miles. Area of upper Lough, 9,453 acres ; of lower Lough, 27,645 acres. These waters are connected by the Ulster Canal with Loughs, Neagh and Belfast.

ERRIGAL.—Is a parish of County Londonderry, Ulster, 5 miles West Northwest of Kilrea. Area, 19,625 acres. Population, 5,784. KERROGUE is also the name of a parish of County Tyrone, Ulster. Area, 21,139 acres. Population, 9,171. TROUGH is also the name of a parish of Counties, Monaghan and Tyrone, Ulster. Area, 25,000 acres. Population, 9,585.

ERRIS.—Is a maritime district or barony of County Mayo, Connaught. Area, 232,889 acres. Population, 26,428. The scenery is very wild and desolate. ERRISHEAD is a lofty promontory in this district, forming the West point of the bay of Broadhaven, 5½ miles North of Belmullet.

ESK.—Is a Lough of County Donegal, Ulster, 3 miles North Northeast of Donegal. Area, 976 acres. It is very beautiful.

ESK.—Is a mountain range between the Counties of Cork and Kerry, Munster.

EYRECOURT.—(or Aircourt.) Is a small market town of County Galway, Connaught, 5 miles Northwest of Banagher. Population, 1,419. The residence of the Eyre family is in this vicinity.

FAHAN.—There are two parishes by this name. One in County Donegal, Ulster, (or lower,) is on Lough Swilly. Area, mostly mountainous, 24,782 acres. Population, 5,823. Another (or upper) 2½ miles South of Buncrana. Area, 10,040 acres. Population, 2,949.

FAIR-HEAD.—(or Benmore Head.) Is a promontory on the North coast of County Antrim, Ulster, 5½ miles East Northeast of Ballycastle. It is 530 feet in height.

FANLOBBUS.—Is a parish of County Cork, Munster. Area, 35,606 acres. Population, 12,253. It comprises the town of Dunmanway.

FANNET-POINT.—Is a headland of County Donegal, Ulster. It has a Light-house on it and is on the West side of the entrance to Lough Swilly.

FAUGHANVALE.—Is a parish of County Londonderry, Ulster, 6 miles East Northeast of Londonderry. Area, 18,582 acres. Population, 5,929.

FEACLE.—Is a parish of County Clare, Munster, 5 miles West Northwest of Scariff. Area, 36,972 acres. Population, 10,156.

FEALE.—Is a river of Counties, Cork, Limerick and Kerry, Munster. It flows Northwest and joins the Shannon river near Guisborough by a navigable branch called the Cashen. Length 30 miles.

FENAGH.—There are two parishes by this name. One is in County Leitrim, Connaught, 2½ miles South Southwest of Ballinamore. Area about 9,800 acres. Population, 4,426. Another is in County Carlow, Leinster, 5 miles East Southeast of Leighlin Bridge. Population, 4,314.

FENNIT.—Is an island off the coast of County Kerry, Munster, forming the division between Tralee and Ballyheigue Bays, 8 miles West Northwest of Tralee. Area, 686 acres. Population, 215.

FERBANE.—Is a village of County Kings, Leinster, on the Brosna river, 9 miles Northeast of Banagher. Population, 537.

FERGUS.—Is a river of County Clare, Munster, and flows into

the Shannon by a broad estuary after a Southeast course of 27 miles.

FERMANAGH.—Is an inland County of Ulster, bounded by the Counties, Donegal, Tyrone, Monaghan, Cavan, Leitrim and Connaught. Area, 457,195 acres; of which 114,847 acres are waste, and 46,755 acres are under water. The surface is varied and the soil is a rich loam. Farms mostly very small and agriculture very backward except in the Northern portion. Oats, Barley, Wheat, Flax and Potatoes are the principal productions of the soil. Fermanagh is divided into 8 baronies and 18 parishes, 15 of which are in the diocese of Clogher. It sends 3 members to the House of Commons.

FERMOY.—Is a market town and parish of County Cork, Munster, 19 miles North Northeast of Cork, on the right bank of the Blackwater. Population of town in 1851, 5,844. The town is well built and the streets are fine. There is a handsome bridge of 13 arches erected in 1689, Barracks for 3,000 troops, Churches, etc. It has extensive Flour Mills and a large trade in agricultural produce.

FINGAL.—Is a district of County Dublin, Leinster. It gives the title of Earl to the Plunkett family. It was originally settled by emigrants from Finland and Norway, and they still retain a dialect and other marks of their foreign origin.

FINGLASS.—Is a parish of County Dublin, Leinster, 4 miles North Northwest of Dublin. Area, 4,696 acres. Population, 2,187. Fair, May 6; the ancient "May Sports" which attract great numbers from Dublin. The poet Parnell was Vicar of Finglass.

FINTONA.—Is a market town of County Tyrone, Ulster, 8 miles Northwest of Clongher. Population, 1,327.

FINVOY.—Is a parish of County Antrim, Ulster, 4 miles South Southwest of Ballymoney. Area, 16,474 acres. Population, 6,405.

FLISK.—Is a small river which flows into the Lakes of Killarney.

FORE.—Is a decayed town of County Westmeath, Leinster, 3 miles East of Castle-Pollard. Population, 119.

FORKHILL.—Is a parish of County Armagh, Ulster, 8 miles South Southwest of Newry. Area, 12,600 acres. Population, 8,128.

FORTH MOUNTAINS.—Is a range of hills in County Wexford, Leinster, 4 miles West of Wexford.

FOXFORD.—Is a small market town of County Mayo, Connaught, 9 miles South of Ballina. Population, 680. It has a Market House and Barracks.

FOYLE.—Is a river of the province of Ulster, rises at Lifford by the junction of the rivers Finn and Mourne, flows North for 14 miles and empties into Lough Foyle. It is navigable for vessels of 600 tons to Londonderry and has a large Salmon fishery.

FOYLE.—(LOUGH) Is a small arm of the sea forming the estuary of the above river, 18 miles in length.

FRANKFORD.—Is a market town of County Kings, Leinster, 8¼ miles Northeast of Birr. Population, 1,345.

FRANKFORT.—Is a desmesne of County Kilkenny, Leinster, 3¾ miles Northeast of Ullingford. It gives the title of Baron to the Montmorency family, descendants of the nephew of Earl Strongbow.

FRESHFORD.—Is a market town of County Kilkenny, 9 miles North Northwest of Kilkenny. Population, 2,075. It has an ancient Church, formerly part of an Abbey.

FUERTY.—Is a parish of County Roscommon, Connaught, 3½ miles West Southwest of Roscommon. Area, 13,375 acres. Population, 5,810.

GALBALLY.—Is a parish of County Limerick, Munster, 7 miles South Southwest of Tipperary. Area, 15,457 acres. Population, 6,651.

GALLOON.—Is a parish of County Fermanagh, Ulster. Area about 25,000 acres. Population, 11,135. It comprises the village of Newton Butler.

GALLYHEAD.—Is a promontory of County Cork, Munster, between Ross and Clonakilty bays.

GALTEE MOUNTAINS.—Are a range of mountains extending from East to West for about 20 miles, between Cahir in County Tipperary, and Charlesville in County Limerick, Munster; some of them are over 2,000 feet in height.

GALWAY.—Is a maritime County of Connaught, having the Counties, Mayo and Roscommon on the North, Counties, Roscommon, Kings and Tipperary on the East, County Clare and the Bay of Galway on the South, and on the West the Atlantic Ocean. Area, 1,565,726 acres, of which about 800,000 acres are waste and water. Population in 1851, 298,129. The surface in the East is flat and fertile, interspersed with bogs, but in the West and on the coast is mountainous and rocky. The climate is good, being mild and humid. The Shannon, Black and Suck are the principal rivers. Agriculture is very backward, the land being better adapted for grazing, and the breed of long-horned cattle is much esteemed. Average rent of land, 12s. 1d. per acre. The fisheries are very valuable. The County, which is second in size in Ireland, is divided into 2 Ridings, East and West, and comprises 16 baronies and 116 parishes in the diocese of Clonfert, Tuam, Kilmacduagh, Elphin and Killaloe. This County sends four members to the House of Commons. The chief towns are Galway, Tuam and Ballinsaloe.

GALWAY.—Is a parliamentary and municipal borough, seaport and market town of County Galway, Connaught, situated at the mouth of the river flowing from Lough Corrib into Galway bay, 105 miles West of Galway. Latitude, 53° 15′ North. Longitude, 9° 3′ West. Area of borough, 628 acres. Population, 24,697. It is very poorly built and the streets are narrow and dirty. It has two bridges, some remains of ancient fortresses, Church, Roman Catholic Cathedral, Grammar School, Court House, County Jail, etc. The harbor has a Light-house and good docks, capable of admitting vessels of 500 tons burthen. Markets, Wednesday and Saturday. The borough sends two members to the House of Commons. It gives title of Viscount to a branch of the Arundel family.

GALWAY BAY.—Is a large inlet of the Atlantic Ocean, between Counties, Galway and Clare, Connaught and Munster. Length, 30 miles. Breadth about 10 miles. Galway is the only town of importance on its shores.

GARTAN.—Is a parish of County Donegal, Ulster, 7 miles North Northwest of Letter-Kenny. Area, 44,124 acres. Population, 2,082.

GARTAN.—(Lough) Is about 2 miles in length.

GARVAGH.—Is a small market town of County Londonderry, Ulster, 9 miles South of Coleraine. Population, 851. It gives the title of baron to the Cannuig family, who own the town and whose seat is adjacent.

GAROAGHY.—Is a parish of County Down, Ulster. 4 miles South Southeast of Dromore. Area, 10.000 acres. Population, 5,063.

GIANTS CAUSEWAY.—Is the celebrated basattic formation

on the North coast of County Antrim, Ulster, 2 miles North Northeast of Bushmills, to the West of Bengore Head. It is a platform projecting into the sea, from the base of a stratified cliff, about 400 feet in height. It is separated into 3 parts comprising together about 40,000 perfectly formed, closely united and very dark colored polygonal columns, each consisting of several pieces, the joints of which are articulated with the greatest nicety. Popular legend ascribe this formation to the labor of Giants, seeking to construct a road across the sea to Scotland.

GLANMIRE.—Is a village of County Cork, Munster, 5 miles East Northeast of Cork. Population, 453.

GLANWORTH.—Is a parish of County Cork, Munster, 5½ miles South Southwest of Mitchelstown. Area, 9,700 acres. Population, 4,832; of village, 1,012.

GLASNEVEN.—Is a parish of County Dublin, Leinster, 3¼ miles North Northwest of Dublin. It was until recently a favorite place of residence, and it has a Botanic Garden belonging to the Dublin Royal Society.

GLASSLOUGH.—Is a small market town of County Monaghan, Ulster, 6 miles East Northeast of Monaghan Population, 562. Adjoining it is Leslie Castle.

GLASSTOOLE.—Is a village of County Dublin, Leinster, 5 miles Southeast of Dublin, on Dublin bay. Population, 849.

GLENA.—Is a beautiful vale and bay of County Kerry, Munster, near Killarney. Lord Kenmore has a cottage here.

GLENAOY.—Is a parish of County Antrim, Ulster, near Lough Neagh, 9 miles Northwest of Lisburn. Population, 3,773.

GLENARM.—Is a market town of County Antrim, Ulster, on an inlet of the Irish sea, 25½ miles North of Belfast. Population, 881.

GLENBEGH.—Is a parish of County Kerry, Munster, on Dingle bay, 13 miles Northeast of Cahirciveen. Area, 30,808 acres. Population, 3,011.

GLENCOLUMBKILL.—Is a parish of County Donegal, Ulster, 12 miles West Northwest of Killybegs. Area, 32,243 acres. Population, 4,356.

GLENDERMOT.—(or CLONDERMOT.) Is a parish of County Londonderry, Ulster. Area, 21,508 acres. Population, 10,295.

GLENDALOUGH.—Is a lough and valley of County Wicklow, Leinster, 24 miles South of Dublin. Famous for its wild grandeur and picturesque ruins.

GLENGAD.—Is a headland of County Donegal, Ulster, 8½ miles East Southeast of Marlin Head. It forms the West point of Culdaff bay.

GLENGARIFF HARBOR.—Is a branch of Bantry bay, County Cork, Munster, 5 miles Northwest of Bantry, on the North side of the bay.

GLENMALURE.—Is a wild mountain vale of County Wicklow, Leinster, on the Avonbeg river. It was the scene of outrages during the rebellion of 1798.

GLENTIES.—Is a village of County Donegal, Ulster, 6 miles East Northeast of Adara. Population, 317. It has an Inn for tourists.

GLIN.—Is a market town and seaport of County Limerick, Munster, on the Shannon river, 18 miles Northeast of Tralee. Population, 1,208. Here is the Castle of the Knights of Glinn, descendants of the Desmond family.

GOLDEN BRIDGE.—Is a village of County Dublin, Leinster, 1¾ miles West Southwest of Dublin. Population, 1,090.

GOLDEN.—Is a small market town of County Tipperary, Munster. In "the Golden Vale," a rich valley of the Suir river, 3½ miles West of Cashel. Population, 602.

GORESBRIDGE.—Is a village of County Kilkenny, Leinster, 2¾ miles East of Gowran. Population, 921.

GOREY.—Is a municipal borough and market town of County Wexford, Leinster, 24 miles North Northeast of Wexford. Population, 3,365. It is neatly built and has a handsome Church, Barracks and Market-House.

GORMANSTOWN.—Is a village of County Meath, Leinster, 1¾ miles Northwest of Balbriggan. Population, 160. It gives the title of Viscount to the Preston family.

GORT.—Is a market town of County Galway, Connaught, 16 miles North Northeast of Ennis. Population, 3,056. It is very neatly built and completely hidden amongst trees. It gives the title of Viscount to the Vereker family, whose mansion, Loughcooter, is two miles from the town.

GRACEHILL.—(or BALLYKENNEDY.) Is a Moravian settlement of County Antrim, Ulster, 2 miles West Southwest of Ballymena Population, 297.

GRAIGUE.—Is a town of County Queens, Leinster, a suburb of Carlow. Population, 1,675. Is also the name of a town of County Kilkenny, Leinster, on the Barrow river, 5 miles West of Goresbridge. Population, 2,448. It has the ruins of a Castle and Abbey.

GRANARD.—Is a market town and parish of County Longford, Leinster, 59 miles West Northwest of Dublin. Area of parish, 18,000 acres. Population, 10,193 ; of town, 2,408. It is neatly built with usual public buildings and has some manufactories. It gives the title of Earl to the Forbes family.

GRAND CANAL.—Is in Counties, Dublin, Kildare and Kings, proceeds from Dublin Westward and joins the Shannon near Banagher. Length, 85 miles. Breadth at surface, 40 feet. Depth, 6 feet. It has a branch 27 miles to Athy where it joins the Barrow river, also branches in Ballinasloe, Portarlington, Mountmellick, etc. It was begun in 1765 and cost to complete £2,000,000. Annual amount of tolls £40,000.

GRANGE.—Is a parish of County Armagh. Ulster, 2½ miles North of Armagh. Area. 6,800 acres. Population, 3,823. There are several smaller parishes with the same name.

GREEN CASTLE.—Is a fort, harbor, coast-guard, pilot and fishing station of County Donegal, Ulster, at the West entrance to Lough Foyle, 4 miles Northeast of Moville. Is also the name of a village of County Down, Ulster, on the North side of Carlingford, bay, 1¾ miles West Northwest of Cranfield point.

GRENAUGH.—Is a parish of County Cork, Munster, 4½ miles North of Blarney. Area, 13,558 acres. Population, 5,351.

GREY ABBEY.—Is a parish of County Down, Ulster, on Lake Strangford, 7 miles South Southwest of Donaghadee. Area, 7,689 acres. Population, 745. Mount Stewart, seat of the Marquis of Londonderry, is here.

GREYSTONES.—Is a headland, fishing village and coast-guard station of County Wicklow, Leinster, 3 miles Southeast of Bray.

GUIBARRA.—Is a small river of County Donegal, Ulster, flowing into an inlet by the same name, between the bays of Rosmore and Trawenagh, after a Southwest course of 13 miles.

GUIBARRA BAY.—Is 5 miles South of Dunsloe.

HACKETSTOWN.—Is a parish and town of Counties, Carlow and Wicklow, Leinster, 8 miles Southeast of Baltinglass. Area, 11,616 acres. Population, 4,223; of town 1,021.

HARRISTOWN.—Is a village of County Kildare, Leinster, 2½ miles Northeast of Kilcullen Bridge. Is also the name of a parish of County Kildare, Leinster, on the river Barrow, 4 miles South Southwest of Kildare. Area, 4,680 acres. Population, 920.

HARROLD'S-CROSS.—Is a village of County Dublin, Leinster, 1 mile South of Dublin Castle. Population, 2,789.

HAWLBOWLINE.—Is an island in Cork harbor, County Cork, Munster, ¾ miles South of the Cove. Is also the name of a rock of County Down, Ulster, off the entrance to Carlingford harbor.

HEADFORD.— Is a market town of County Galway, Connaught, 9 miles Southwest of Tuam. Population, 1,647. It is neatly built and adjoining is the mansion of the St. George family.

HILLSBOROUGH.—Is a market town and parish of County Down, Ulster, 3 miles South Southwest of Lisburn. Population of parish, 6,524; of town, 1,338. Market, Thursday. The seat of the Marquis of Downshire is adjacent.

HOLLYMOUNT.—Is a town of County Mayo, Connaught, on the Robe river, 4½ miles East Northeast of Ballinrobe. Population, 454.

HOLLYWOOD.—Is a village and parish of County Down, Ulster, on Belfast Lough, 4½ miles Northeast of Belfast. Population of village, 1,532. It is remarkably well built, and in its vicinity are many handsome mansions and villas occupied by merchants from Belfast, whose families resort here for sea-bathing. Is also the name of a parish of County Dublin, Leinster, 2½ miles East Southeast of Naul. Area, 3,992 acres. Population 1,022. Is also the name of a parish of County Wicklow, Leinster, 2½ miles Southeast of Ballymore-Eustace. Area of parish, 18,383 acres. Population, 2,770. Its village is miserable.

HORE-ABBEY.—Is a parish of County Tipperary, Munster. Population, 536.

HORSE-ISLAND.—Is an island at the West side of the entrance of Castle-Townsend haven, County Cork, Munster. It has a landmark Tower.

HOSPITAL.—Is a village and parish of County Limerick, Munster, 11 miles West of Tipperary. Population of parish, 2,538.

HOWTH.—(THE HILL OF) Is a peninsular and parish of County Dublin, Leinster. Area of parish, 2,760 acres. Population, 1,538 ; mostly engaged in fisheries, and of the village, (8 miles by Railroad East Northeast of Dublin,) 692. It has a large Harbor of refuge, extensive Docks, Light-house, etc., constructed at a cost of £2,000,000, but from its position, the rocks which still obstruct it, and the accumulation of sand, is now almost useless ; at the extremity of the peninsular is a handsome Light-house with a fixed red light. Howth gives the title of Earl to the St. Lawrence or Tristram family.

HUNGOY HILL.—Is a mountain of County Cork, Munster, 15 miles West Northwest of Bantry. It is 2,249 feet in height. From a lake on its summit descends a torrent in broken cascades, one of which, 700 feet in height, is said to be the finest in the Kingdom.

ICHTERMURROCH.—Is a parish of County Cork, Munster, 2½ miles East Southeast of Castle-Martyr. Area, 5,556 acres. Population, 3,092.

IMOGEELY.—Is a parish of County Cork, Munster. Area, 6,430 acres. Population, 3,121.

INCH.—There are several parishes by this name, as follows, viz : County Down, Ulster, 2½ miles North of Downpatrick. Area, 6,494 acres. Population, 2,489 ; Counties, Wicklow and Wexford, Leinster, 2¾ miles Southwest of Arklow. Area, 5,943 acres. Population, 2,006 ; County Cork, Munster, 5½ miles Southwest of Cloyne. Area, 3,823 acres. Population, 1,617 ; County Donegal, Ulster, 1 mile West of Churchtown. Area, 3,100 acres. Population, 978. It comprises the island of Inch in Lough Swilly; County Wexford, Leinster, 6½ miles West Southwest of Taghmon. Area, 1,389 acres. Population, 526.

INCHICRONANE.—(or INNISCRONANE.) Is a parish of County Clare, Munster, 5¼ miles North Northeast of Ennis. Area, 17,438 acres. Population, 5,118.

INCHEGEELAGH.—(or EVELEARY.) Is a parish of County Cork, Munster, 9 miles West Southwest of Macroone. Area, 45,415 acres. Population, 6,357 ; of village, 233.

INCHIQUIN.—Is a barony and island of County Clare, Munster. The island is situated in Lough Corrib ; the barony has an area of 83,387 acres. Population, 21,231. It contains the ruins of Inchiquin Castle, also Lough Inchiquin.

INNISBOFFIN.—(or BOFFIN) Is a parish of County Mayo, Connaught, 3 miles North Northwest of Claggan Point. It comprises an island of the same name. Area, 3,152 acres. Population, 1,612. It has a good harbor on its South coast.

INNISBOFFIN.—Is the name of several islands in Counties, Donegal and Longford.

INNISCALTHRA.—Is a parish of Counties, Clare and Galway, Munster and Connaught, 3¼ miles East Northeast of Scariff. Area, 11,284 acres. Population, 2,378. Holy Island is in this parish.

INNISCARRA.—Is a parish of County Cork, Munster, 5½ miles West Southwest of Cork. Area, 10,190 acres. Population, 4,407. There is also a small island of the same name in County Donegal, Ulster, 1¼ miles South of Arran.

INNISCATTERY.—Is an island of County Clare, Munster, in an estuary of the river Shannon, 2 miles South Southwest of Kilrush. Area, 100 acres. It was formerly a stronghold of the Danes and is in a great part covered with ruins.

INNISHANNON.—Is a decayed inland town and parish of County Cork, Munster, 12 miles South Southwest of Cork, on the river Bandon. Area, 7,153 acres. Population, 3,615 ; of town 625.

INNISHARGIE.—Is a parish of County Down, Ulster. Area, 5,516 acres. Population, 3,014. It comprises the town of Kirkcubbin.

INNISSHARK.—Is an island of County Mayo, Connaught, Southwest of Innisboffin. Population, 200.

INNISHERE.—Is an island and parish of County Galway, Connaught. Area, 1,400 acres. Population, 406.

INNISHERKIN.—(or SHERKIN) Is an island at the entrance to Baltimore Bay, County Cork, Munster. Population, 1,026.

INNISKEA.—There are two islands by this name off the West coast of County Mayo, Connaught, 10 miles North Northeast of Achil-Head.

INNISKEEL.—(or INISHKEEL.) Is a maritime parish of County Donegal, Ulster, 11 miles North of Killybegs. Area, 102,082 acres. Population, 12,606. There is also a small island by the same name in Guibarry Bay, County Donegal, Ulster.

INNISMACSAINT.—(or ENNISMACSAINT.) Is a parish of Counties, Fermanagh and Donegal, Ulster. Area, 52,994 acres. Population, 14,693. It comprises a part of the town of Ballyshannon.

INNISMAGRATH.—Is a parish of County Leitrim, Connaught, 5 miles Southeast of Dromahaire. Area, 27,439 acres, including part of Lough Allen. Population, 9,603.

INNISTIOGUE.—Is a disfranchised parliamentary borough, market town and parish of County Kilkenny, Leinster, on the Nore river, here crossed by a handsome bridge of 10 arches, 8 miles West Northwest of New Ross. Area of parish, 9,741 acres. Population, 3,501 ; of town, 956.

INNISTURK ISLAND.—Is an island of the West coast of County Mayo, Connaught, 4½ miles North Northeast of Innisboffin, Population, 500.

INNY.—Is a river rising in Lough Sheelan and flowing Southwest through Counties, Westmeath and Longford, Leinster, falls into Lough Ree.

IRVINESTOWN.—(or LOWTHERSTOWN.) Is a town of County Fermanagh, Ulster, 9 miles North of Enniskillen. Population, 1,388. There are twelve Fairs held here Annually.

ISLANDBRIDGE.—Is a village of County Dublin, Leinster, on the Liffey river, 1¾ miles West of Dublin Castle. Population, 767.

ISLANDEADY.—(or ISLANDINE.) Is a parish of County Mayo, Connaught, 3¾ miles West Northwest of Castlebar. Area, 24,940 acres, including Loughs. Population, 8,463.

ISLAND MAGEE.—Is a parish of County Antrim, Ulster, 7 miles North Northeast of Carrickfergus. Area, 7,037 acres. Population, 2,782.

JAMES.—(ST.) Is a parish of County Wexford, Leinster, on Waterford harbor, 5½ miles North Northwest of Fethard. Area, 8,489 acres ; including Dunbrod and Rathroe. Population, 3,693. There is another parish by this name in County Dublin, Leinster. Area, 1,974 acres. Population, 12,466. It comprises a part of the city of Dublin.

JAMESTOWN.—Is a village of County Leitrim, Connaught, on the Shannon river, 2½ miles Southeast of Carrick. It was formerly a walled town. Population, 315.

JOHN.—(ST.) There are numerous parishes by this name, as follows, viz : County Sligo, Connaught. Area, 7,256 acres. Population, 13,299. It comprises part of the town of Sligo ; County Limerick, Munster. Area, 134 acres. Population, 12,775. It comprises part of the city of Limerick ; County Kilkenny, Leinster,

Area, 5,532 acres. Population, 5,448. It comprises part of the city of Kilkenny; County Dublin, Leinster. Area, 14 acres. Population, 3,931. It is wholly comprised in the city of Dublin; County Waterford, Munster. Area, 732 acres. Population, 3,313. It comprises part of the city of Waterford; County Waterford, Munster. Area, 13 acres. Population, 3,166. It is wholly included in the city of Waterford; County Wexford, Leinster. Area, 525 acres. Population, 2,954. It is comprised in the town of Wexford; County Roscommon, Connaught, 9 miles North Northwest of Athlone. Area, 11,635 acres. Population, 2,806; County Kildare, Leinster. Area, 1,123 acres. Population, 1,781. It comprises part of the town of Athy; County Wexford, Leinster. Area, 2,207 acres. Population, 675. Other parishes comprise portions of the towns of Sligo, Wexford, Kilkenny, Dublin, Waterford and Limerick.

KANTURK.—Is a market town of County Cork, Munster, 11 miles West Southwest of Buttevant. Population, 4,388. It contains some good public buildings, and gives the title of Viscount to Earl of Egmont, whose residence is in the village.

KELLS.—Is a municipal borough, market town and parish of County Meath, Leinster, on the Blackwater river, 36 miles Northwest of Dublin. Area of parish, 8,597 acres. Population, 7,648; of town, 4,205. It is pleasantly situated and has a handsome Church and other good buildings. Is also the name of a parish of County Kilkenny, Leinster, 8 miles South of Kilkenny. Population, 1,831. It has the ruins of an Abbey of the twelfth Century. Is also the name of a coast-guard and fishing station of County Kerry, Munster, on Dingle Bay.

KENMARE.—Is a market town and parish of County Kerry, Munster, 13 miles South Southwest of Killarney. Area of parish, 22,490 acres. Population, 5,839; of town, 1,339. It is isolated and surrounded by picturesque scenery. It has an elegant suspension across the estuary of the Roughty and a convenient harbor and pier approached by vessels of large burthen. It gives the title of Earl to the Brown family.

KERRY.—Is a maritime County of Munster. On the North is the estuary of the Shannon river, on the East and South are Counties, Limerick and Cork, and on the West the Atlantic Ocean. Area, 1,186,126 acres; of which 400,000 acres are waste. Population, 238,239. The surface is wild, rugged and mountainous, the

Macgillicuddy Reeks, the loftiest mountains in Ireland, being in this County. The coast is deeply indented with bays, of which Tralee, Dingle and Kenmare are the principal ones. Dunmore Head, between Tralee and Dingle bays, is the most Westerly land in Ireland. The Feale, Maine, Laune or Lane, and Roughty are the chief rivers. Chief Loughs: Killarney, Carra and Currane. The soil is very poor, and the climate, except on the seaboard, is usually mild. Agriculture is very backward, Potatoes, Wheat and Barley being the principal productions. The fisheries in 1836 employed 6,311 men. Coal, Iron, Copper, Lead and Slate are also found. The manufactures are unimportant. In 1834 there were about 20,000 scholars attending school in this County, of whom 19,000 were Roman Catholics. Kerry is divided into 8 baronies and 83 parishes in the diocese of Ardfert. The County sends two members to the House of Commons and the borough of Tralee one member. It was made a *Shire* in 1210 by King John, and gives the title of Earl to the Marquis of Lansdowne.

KERRY HEAD.—Is a promontory of County Kerry, Munster, South of the entrance to Shannnon.

KILBALLYHONE.—Is a parish of County Clare, Munster, 13 miles West Southwest of Kilrush. Area, 10,835 acres Population, 4,346.

KILBARRON.—Is a parish of County Donegal, Ulster. Area, 23,932. Population, 10,027. It includes part of the town of Ballyshannon.

KILBEAGH.—Is a parish of County Mayo, Connaught, 8 miles West Northwest of Ballaghadireen. Area, 33,824 acres. Population, 9,963.

KILBEGGAN.—Is a parish and market town of County Westmeath, Leinster, on the upper Brosna river and a branch of the Grand Canal, 6½ miles North of Tullamore. Population, 1,910.

KILBEGNOT.—Is a parish of County Galway, Connaught, 6¼ miles North of Ballinamore. Area, 10,867 acres. Population, 5,036.

KILBEHENNY.—Is a parish of County Limerick, Munster, 4 miles East Northeast of Mitchelstown. Area, 15,376 acres. Population, 4,291.

KILBOLANE.—Is a parish of County Cork, Munster, 9 miles West Southwest of Charleville. Area, 10,015 acres. Population, 4,155.

KILBRIDE.—There are several parishes by this name, as follows, viz : Counties Cavan and Meath, Ulster and Leinster, 3 miles West Northwest of Oldcastle. Area, 9,341 acres. Population, 5,041 ; County Roscommon, Connaught, 5 miles West of Roscommon. Area, 19,287 acres. Population, 8,578 ; County Kings, Leinster. Area, 10,152 acres. Population, 9,608. It comprises the town of Tullamore, and has the ruins of six Castles.

KILBRIN.—Is a parish of County Cork, Munster, 9½ miles Northwest of Mallow. Area, 12,631 acres. Population, 4,855.

KILBROGAN.—Is a parish of County Cork, Munster. Area, 7,578 acres. Population, 5,404. It comprises a part of the town of Bandon.

KILCAR.—Is a parish of County Donegal, Ulster, on Donegal bay, 5 miles West of Killybegs. Area, 18,883 acres. Population, 4,969.

KILCASKIN.—Is a parish of Counties, Cork and Kerry, Munster, 8½ miles East Northeast of Castletown-Berehaven. Area, 51,491 acres. Population, 6,780.

KILCATERN.—Is a parish of County Cork, Munster, 7¼ miles North Northwest of Castletown-Berehaven. Area, 21,778 acres. Population, 6,940.

KILCLOONEY.—Is a parish of County Armagh, Ulster. Area, 12,833 acres. Population, 8,079. It comprises part of the town of Markethill.

KILCOCK.—Is a market town of County Kildare, Leinster, on the Grand Canal, 3 miles West Northwest of Maynooth. Population, 1,537

KILCOLEMAN.—There are several parishes with this name, as follows, viz : County Kerry, Munster. Area, 7,758 acres. Population, 4,745. It comprises the town of Milltown ; County Mayo, Connaught. Area, 23,739 acres. Population, 9,451. It comprises the town of Clare-Morris.

KILCOLEMAN.—Is a ruined Castle of County Cork, Munster, 2 miles North of Doneraile. It was the domain of the poet Spencer, and the place where he composed the greater part of his "Fairy Queen."

KILCOMMON.—Is a parish of County Mayo, Connaught. Area, 203,396 acres. Population, 17,000. It comprises the village of Belmullet and forms the principal part of the wild mountain land of Eoris. Is also the name of a parish of County Mayo,

Connaught, 4½ miles East Northeast of Ballinrobe. Area, 17,395 acres. Population, 7,156.

KILCONDUFF.—Is a parish in County Mayo, Connaught. Area, 16,522 acres. Population, 7,072. It comprises the town of Swineford.

KILCONNEL.—Is a parish of County Galway, Connaught, 7¼ miles West Northwest of Ballinasloe. Area, 6,082 acres. Population, 1,880.

KILCOO.—Is a parish of County Down, Ulster. Area, 18,205 acres. Population, 6,583. It includes the town of Newcastle.

KILCOOLEY.—Is a parish of Counties, Tipperary and Kilkenny, Munster and Leinster. Area, 11,500 acres. Population, 4,006. It comprises the town of New Birmingham. There are others of same name in Counties, Meath, Roscommon and Galway.

KILCOOHANE.—There are two parishes by this name in Munster. One is in County Kerry, 17 miles West Southwest of Kenmare. Area, 63,702 acres. Population, 10,776; another is in County Cork, 13 miles West Southwest of Bantry. Area, 14,588 acres. Population, 4,856.

KILCRONAGHAN.—Is a parish of County Londonderry, Ulster. Area, 7,792 acres. Population, 4,245. It comprises the town of Tubbermore.

KILCULLEN-BRIDGE.—Is a small town of County Kildare, Leinster, on the river Liffey, 5 miles South Southwest of Naas. Population, 2,442.

KILCUMMIN.—There are two parishes by this name. One is in County Galway, Connaught. Area, 108,791 acres. Population, 10,824. It comprises the town of Cughterard; another is in County Kerry, Munster, 4 miles West Northwest of Killarney. Area, 38,953 acres. Population, 7,360.

KILDALLON.—Is a parish of County Cavan, Ulster, 3½ miles North Northwest of Killeshandra. Area, 11,986 acres. Population, 4,480.

KILDRESS.—Is a parish of County Tyrone, Ulster, 3½ miles North Northwest of Cookstown. Area, 26,251 acres. Population, 8,192.

KILDRUMFERTON.—Is a parish of County Cavan, Ulster, 6 miles West Southwest of Ballinanagh. Area, 16,400 acres. Population, 10,446.

KILDYSERT.—Is a parish of County Clare, Munster, 12 miles

South Southwest of Ennis. Area, 12,859 acres. Population, 5,130. It consists partly of islands in the estuary of the Fergus and Shannon rivers.

KILFARBOY.—Is a parish of County Clare, Munster. Area, 13,981 acres. Population, 7,498. It comprises the town of Miltown-Malbay.

KILFEDANE.—Is a parish of County Clare, Munster, 4½ miles West Southwest of Kildysert. Area, 13,733 acres. Population, 4,661.

KILFENNORA.—Is a parish of County Clare, Munster, 5 miles Northeast of Ennistymon. Area of parish, 10,777 acres. Population, 3,206.

KILFERGUS.—Is a parish of County Limerick, Munster. Area, 14,207 acres. Population, 5,052. It comprises the town of Glin.

KILFIERAGH.—Is a parish of County Clare, Munster, on the Atlantic, 7½ miles West Northwest of Kilrush. Area, 9,870 acres. Population, 7,137.

KILFINANE.—Is a parish and village of County Limerick, Munster, 5 miles Southeast of Kilmallock. Area, 6,487 acres. Population, 4,356 ; of village, 1,782. Near it are the ruins of Castle Oliver.

KILFREE.—Is a parish of County Sligo, Connaught, 10 miles West Southwest of Ballinafad. Area, 14,313 acres. Population, 6,048.

KILFYAN.—Is a parish of County Mayo, Connaught, 6¼ miles West of Killalla. Area, 28,735 acres. Population, 6,040.

KILGARRIFFE.—Is a parish of County Cork, Munster. Area, 4,328 acres. Population, 6,432. It comprises the town of Clonakilty.

KILGAROEN.—Is parish of County Kerry, Munster, 7 miles East Northeast of Kenmare. Area, 43,631 acres, mostly mountainous. Population, 3,988.

KILGARVEY.—Is a parish of County Mayo, Connaught, 5 miles East Southeast of Ballina. Area, 19,879 acres. Population, 4,158.

KILGEEVER.—Is a parish of County Mayo, Connaught, 11 miles West Southwest of Westport. Area, 58,089 acres. Population, 12,583.

KILGLASS.—There are several parishes by this name, as follows, viz : County Roscommon,*Connaught, 5 miles Northeast of Strokestown. Area, 15,970 acres. Population, 10,053 ; County Sligo, Connaught, 4¾ miles Southwest of Easkey. Area, 12,884 acres. Population, 4,941 ; County Longford, Leinster, 3½ miles Southwest of Edgeworthstown. Population, 2,977 ; including Ahara.

KILKEE.—Is a small town of County Clare, Munster, on a bay of the same name, 8 miles West Northwest of Kilrush. Population, 1,481.

KILKEEDY.—There are two parishes by this name. One is in County Clare, Munster, 5 miles North Northeast of Corrofin. Area, 18,629 acres. Population, 4,192 ; another is in County Limerick, Munster, 4½ miles Southwest of Limerick. Area, 8,881 acres. Population, 4,109.

KILKEEL.—Is a town and parish of County Down, Ulster, on the Kilkeel river, 1 mile above its mouth, and 7½ miles East Southeast of Rostrevor. Area of parish, 47,882 acres. Population, 16,269 ; of town, 1,146.

KILKEEVEN.—Is a parish of County Roscommon, Connaught. Area, 27,007 acres. Population, 10,922. It comprises the town of Castlereagh.

KILKERRAN BAY.—Is a bay on the West coast of County Galway, Connaught, district of Connemara. There are numerous islands in this bay.

KILLECONNEAGH.—Is a parish of County Cork, Munster. Area, 19,295 acres. Population, 7,085. It comprises the village of Castletown-Berehaven.

KILLAGHTEE.—Is a parish of County Donegal, Ulster, 3 miles East Southeast of Killybegs. Area, 13,368 acres. Population, 5,803.

KILLAGHY.—Is a parish of County Kings, Leinster, 2¾ miles East of Ballyboy. Area, 18,132 acres. Population, 4,421.

KILLALLA.—Is a small seaport town and Bishop's See of County Mayo, Connaught, on a large inlet of the Atlantic of the same name, 7½ miles North Northwest of Ballina. Population, 1,446.

KILLALOE.—Is a thriving market town and Episcopal See of County Clare, Munster, on the Shannon river, here crossed by a bridge of 19 arches, 11 miles North Northeast of Limerick. Population, 2,773, including the suburb of Ballina.

KILLANEY.—Is a small bay and fishing village of County

Galway, Connaught, near the East end of the island of Arranmore. Population of village, 604. It has a harbor and coast guard station. Is also the name of a parish in County Louth, Ulster, 4 miles West of Louth. Population, 4,896.

KILLANIN.—Is a parish of County Galway, Connaught, 8½ miles Northwest of Galway. Area, 71,463 acres. Population, 11,278.

KILLARD.—Is a parish of County Clare, Munster 6½ miles Northeast of Kilkee. Area, 17,022 acres. Population, 6,941.

KILLARE.—Is a parish of County Westmeath, Leinster. Area, 11,281 acres. Population, 4,000. It comprises part of the town of Ballymore.

KILLARGEY.—Is a parish of County Leitrim, Connaught, 5½ miles South Southwest of Manor-Hamilton. Area 14,893 acres. Population, 4,873.

KILLARNEY.—Is a market town and parish of County Kerry, Munster, 44 miles West Northwest of Cork, and 16 miles North Northeast of Kenmare. Area of parish, which includes only a part of the famed lake scenery, 38,151 acres. Population, 10,476; of town, 7,127. It has two or three good streets and many miserable and filthy alleys. It contains a parish church, Roman Catholic Cathedral, etc. The town is mainly supported by tourists, and has several good hotels. Markets, Saturday.

KILLARNEY.—(Lakes or) Are three connected lakes of County Kerry, Munster. The most Southern lake is 3½ miles in length by 2 miles in breadth, and is divided from the middle lake by a projecting peninsula on which stand the picturesque remains of Muckrass Abbey. These lakes are famed for their picturesque beauty and have some handsome cascades and fine woods. They are fed by the Flesk river and many smaller mountain steams, and discharge their superfluous waters at the Northwest extremity of the lower lake by the Laune river.

KILLASHEE.—Is a parish of County Longford, Leinster, containing the village, of Killashee and Cloondara the former on the Royal Canal 4½ miles South Southeast of Farmonbarry. Area, 14,427 acres. Population, 4,491.

KILLASNET.—Is a parish of County Leitrim Connaught Area, 26,918. Population, 6,286. It comprises part of the town of Manor-Hamilton.

KILLASSER.—Is a parish of County Mayo, Connaught, 6 miles

East Northeast of Foxford. Area, 19,677 acres. Population, 6,962.

KILLEAD.—(or KILLEAGH.) Is a parish of County Antrim, Ulster, 5 miles South of Antrim. Area, 42,836 acres. Population, 6,725.

KILLEBAN.—Is a parish of County Queens, Leinster. Area, 25,995 acres. Population, 12,939. It comprises the villages of Ballylinon and Arles.

KILLEDAN.—Is a parish of County Mayo, Connaught, 6 miles West Southwest of Swineford. Area, 14,515 acres. Population, 6,410.

KILLEEDY.—Is a parish of County Limerick, Munster, 5½ miles South of Newcastle. Area, 25,456, acres. Population, 6,341.

KILLEEVAN.—Is a parish of County Monaghan, Ulster. Area, 11,571, acres. Population, 8,417. It comprises the village of New-Bliss.

KILLENAULE.—Is a small town of County Tipperary, Munster, 12¾ miles South Southwest of Urlingford. Population, 1,786. It is very poorly built.

KILLENCARE.—Is a parish of County Cavan, Ulster, 4½ miles West Southwest of Bailieboro. Area, 15,911 acres. Population, 8,126.

KILLENUMERY.—Is a parish of County Leitrim, Connaught, 1½ miles South of Droumahaire. Area, 14,086 acres. Population, 4,065.

KILLERERAN.—Is a parish of County Galway, Connaught, 6½ miles Southeast of Tuam. Area, 14.536, acres. Population. 4,782.

KILLERSHERDING.—Is a parish of County Cavan, Ulster, 2¾ miles Southwest of Cootehill. Area, 16,618 acres. Population, 10,208.

KILLESHANDRA.—Is a market town and parish of County Cavan, Ulster, 22 miles South Southeast of Enniskillen. Area of parish, 22.241 acres. Population, 12,552 ; of town, 1,085. Markets, weekly.

KILLESHER.—Is a parish of County Fermanagh, Ulster, 4 miles North Northwest of Swanliba. Area, 24,936 acres. Population, 5,225.

KILLESHILL.—Is a parish of County Tyrone, Ulster, 4½ miles East Northeast of Ballygawley. Area, 9,839 acres. Population, 4,985.

KILLESHIN.—Is a village and parish of County Queens,

Leinster. Area, 10,905 acres. Population, 5,278; including the town of Graigue.

KILLEVEY.—Is a parish of County Armagh, Ulster, 3 miles West of Newry. Area, 28,174 acres. Population, 17.789.

KILLAIN.—Is a parish of County Galway, Connaught, 3¼ miles Northeast of Mount Bellew. Area, 13,564 acres. Population, 5,671.

KILLIMORE.—Is a parish of County Galway, Connaught, 6 miles North Northwest of Portumna. Area, 9,220 acres. Population, 4,140.

KILLINAGH.—Is a parish of County Cavan, Ulster, 10 miles East Southeast of Manor-Hamilton. Area, 24,783 acres. Population, 6,512.

KILLINANE.—Is a parish of County Kerry, Munster, 3½ miles North Northeast of Cahirciveen. Area, 26,868 acres. Population, 3,569.

KILLINCHY.—Is a parish of County Down, Ulster, 9 miles North of Downpatrick. Area, 13,865 acres. Population, 7,470.

KILLINEY.—Is a maritime parish and village of County Dublin, Leinster, on Killiney bay, between Dalky and Bray Head, 2½ miles South Southeast of Kingston. Area of parish, 1,334 acres. Population, 986. It comprises the village of Cabinteely.

KILLOE.—Is a parish of County Longford, Leinster, 5¼ miles Northeast of Longford. Area, 41,513, acres. Population, 19,477.

KILLORAN.—Is a parish of County Sligo, Connaught, 6¼ miles West Northwest of Ballymote. Area, 1,399 acres. Population, 4,044.

KILLORGLIN.—Is a village and parish of County Kerry, Munster, on the Laune river, 13 miles West Northwest of Killarney. Area of parish 30,000 acres. Population, 8,574; of village, 925.

KILLOUGH.—Is a small seaport town of County Down, Ulster, on a bay of same name, 1½ miles West Southwest of Ardglass. Population, 1,148. Fisheries are important. There are four Fairs held here Annually.

KILLUCAN.—Is a large parish of County Westmeath, Leinster, 8½ miles East of Mullingar. Population, 9,562. The village is neatly built.

KILLURSA.—Is a parish of County Galway, Connaught. Area, 8,877 acres. Population, 4,995. It comprises the town of Headford.

KILLURY.—Is a parish of County Kerry, Munster, 4¼ miles North of Ardfert. Area, 11,090 acres. Population, 6,480.

KILLYBEGS.—Is a small seaport town and parish of County Donegal, Ulster, on an inlet of the Atlantic, 14 miles West of Donegal. Population, 3,290 ; of town, 789.

KILLYLEAGH.—Is a seaport town and parish of County Down, Ulster, on Lough Strangford, 16 miles South Southeast of Belfast. Population, 6,688 ; of town, 1,116. It is well built and has some fine buildings and manufactories.

KILLYMAN.—Is a parish of Counties, Tyrone and Armagh, Ulster, on the Blackwater river, 2½ miles North of Moy. Area, 10,559 acres. Population, 8,220.

KILLYMARD.—Is a parish of County Donegal, Ulster, on the North side of Donegal bay. Area, 28,229 acres. Population, 4,743. It comprises part of the town of Donegal.

KILMACABEA.—Is a parish of County Cork; Munster, on Glandore harbor. Area, 13,757 acres. Population, 6,209,

KILLMACALLANE.—Is a parish of County Sligo, Connaught, 4 miles South Southeast of Coloney. Area, 9,928 acres. Population, 5,098.

KILMACDUAGH.—Is a parish of County Galway, Connaught, 3 miles South Southwest of Gort. Area, 8,804 acres. Population, 4,149. It has the ruins of a Cathedral, Abbey and a Round Tower, which leans even more than the tower of Pisa.

KILMACDUANE.—Is a parish of County Clare, Munster, 7 miles North Northeast of Kilrush. Area, 16,701 acres. Population, 6,762.

KILMACOMOGUE.—Is a parish of County Cork, Munster. Area, 58,835 acres. Population, 16,188. It comprises the town of Bantry.

KILMACREHY.—Is a maritime parish of County Clare, Munster, on the Atlantic. Area, 7,403 acres. Population, 4,264. It comprises the village of Liscanor.

KILMACRENAN.—Is a parish of County Donegal, Ulster, 6 miles North Northwest of Letterkenny. Area, 35,617 acres. Population, 9,343.

KILMACTIGUE.—Is a parish of County Sligo, Connaught, 8 miles West Southwest of Tubber-curry. Area, 32,533 acres. Population, 9,097.

KILMACTHOMAS.—Is a town of County Waterford, Munster, 12 miles West Southwest of Waterford. Population, 1,197.

KILMACTRANEY.—Is a parish of County Sligo, Connaught, 6 miles North Northeast of Boyle. Area, 13,447 acres. Population, 4,604.

KILMAINMORE.—Is a parish of County Mayo, Connaught, 5 miles East Southeast of Ballinrobe. Area, 13,792 acres. Population, 4,877.

KILLMALLOCK.—Is a municipal borough and market town of County Limerick, Munster, 19 miles South of Limerick. Population, 1,408. It contains many ruins and a street consisting of antique stone houses, mostly of the date of James I.

KILMALY.—Is a parish of County Clare, Munster, 5 miles West Southwest of Ennis. Area, 23,936 acres. Population, 4,908.

KILMANAHEEN.—Is a parish of County Clare, Munster. Area, 8,177 acres. Population, 6,436. It comprises the town of Ennistymon.

KILMANMAN.—Is a parish of County Queens, Leinster. Area, 16,848 acres. Population, 4,565. It comprises the village of Clonaslee.

KILMEEDY.—Is a parish of County Limerick, Munster, 7 miles East Southeast of Newcastle. Area, 9,057 acres. Population, 4,739.

KILMEEN.—There are several parishes of this name, as follows, viz: County Cork, Munster, 3 miles West Southwest of Newmarket. Area, 36,710 acres. Population, 10,380; County Cork, Munster, 5 miles North Northwest of Clonakilty. Area, 8,667 acres. Population, 3,736. The ruins of Ballyward Castle are here; County Galway, Connaught, 3½ miles West Northwest of Loughrea. Area, 3,808 acres. Population, 980.

KILMEGAN.—Is a parish of County Down, Ulster. Area, 13,970 acres. Population, 7,467. It comprises the town of Castlewellan.

KILMICHAEL.—There are two parishes by this name in Munster. One is in County Clare, 8¼ miles West Northwest of Kildysert. Area, 18,772 acres. Population, 5,080; another is in County Cork, 5¼ miles South Southwest of Macroom. Area, 20,869 acres. Population, 6,250.

KILMINA.—Is a parish of County Mayo, Connaught, 2½ miles North of Westport. Area, 10,762 acres. Population, 7,876.

KILMOE.—Is a parish of County Cork, Munster, on the coast. Area, 13,974 acres. Population, 7,234.

KILMOILY.—Is a parish of County Kerry, Munster, 3 miles West Northwest of Ardfert. Area, 7,750 acres. Population, 4,459.

KILMORE.—There are several parishes of this name, as follows, viz: County Cavan, Ulster, 2½ miles West Southwest of Cavan. Area, 16,885 acres. Population, 7,250. It is an Episcopal See, and has an ancient Cathedral, Church and modern Episcopal Mansion. The diocese comprises 39 parishes, chiefly in Ulster; County Armagh, Ulster. Area, 17,273 acres. Population, 14,256. It comprises the town of Richhill. A desperate affray which took place here between the Roman Catholics and Protestants in 1795, is said to have given rise to the Orange institutions; County Down, Ulster, 1½ miles East Southeast of Ballinahinch. Area, 12,854 acres. Population, 6,277; County Monaghan, Ulster, 2½ miles West Northwest of Monaghan. Area, with lakes, 8,689 acres. Population, 5,121; County Roscommon, Connaught, on the Shannon river, opposite Jamestown. Area, 9,316 acres. Population, 5,164; County Tipperary, Munster, 4 miles South of Nenagh. Area, 13,535 acres. Population, 5,138. It has several ruined Churches and Castles; (or ERRIS,) is in County Mayo, Connaught, on the Atlantic. Area, 29,492 acres. Population, 9,428. Several other smaller parishes have this name.

KILMOREMOY.—Is a parish of County Sligo, Connaught. Area, 12,331 acres. Population, 13,129. It comprises the town of Ballina.

KILMOVEE.—Is a parish of County Mayo, Connaught, 4 miles West Southwest of Ballaghadireen. Area, 20,756 acres. Population, 5,844.

KILMURRY.—There are several parishes by this name in Munster. One is in County Clare, on Clonderalaw bay. Area, 10,457 acres. Population, 4,332; another, (IBRICKANE,) is in County Clare, 5 miles South Southwest of Miltown-Malbay. Area, 25,857 acres. Population, 10,747. There are two Fairs held here Annually.

KILNAMANAGH.—Is a parish of County Cork, Munster, 9 miles Southwest of Castletown-Berehaven. Area, 13,810 acres. Population, 5,861.

KILNAUGHTEN.—Is a parish of County Kerry, Munster.

Area, 9,164 acres. Population, 5,102. It comprises the town of Tarbert.

KILNEBOY.—(or KILLINABOY.) Is a parish of County Clare, Munster. Area, 17,967 acres. Population, 4,102. It comprises the town of Conofin and has many ruins.

KILREA.—Is a parish and market town of County Londonderry, Ulster, on the Bann river, 13 miles West Northwest of Ballymena. Area of parish, 6,313 acres. Population, 4,277; of town, 1,191, partly engaged in Linen weaving. It has a Market House and Public School built by the Mercers Company of London, to whom the town belongs. Is also the name of a parish of County Kilkenny, Leinster, 3 miles Northwest of Knocktopher. Population, 720. One of the finest round Towers in Ireland and a ruined Abbey founded in 1176, are here.

KILRONAN.—There are several parishes by this name, as follows, viz: County Roscommon, Connaught, 9½ miles North Northwest of Carrick-on-Shannon. Area, 16,356 acres. Population, 7,085; County Waterford, Munster, 3½ miles South Southwest of Clonmel. Area, 16.701 acres. Population, 4,772; County Waterford, Munster, 3½ miles Southwest of Waterford. Population, 126.

KILRUSH.—Is a seaport, market town and parish of County Clare, Munster, on an inlet in the estuary of the river Shannon, 27 miles Southwest of Ennis. Area of parish, 15,658 acres. Population, 11,385; of town, 5,071. It is finely built and has manufactories for flannel, frieze, linen sheetings, etc. Market, Saturday.

KILSEILY.—Is a parish of County Clare, Munster. Area, 11,102 acres. Population, 4,469. It comprises the town of Broadford.

KILSHANIG.—Is a parish of County Cork, Munster, 3 miles Southwest of Mallow. Area, 27,595 acres. Population, 9,348.

KILSKEERY.—Is a parish of County Tyrone, Ulster. Area, 20,438 acres. Population, 9,352. It comprises the town of Trillick.

KILSKYRE.—Is a parish of County Meath, Leinster. Area, 11,724 acres. Population, 5,014. It comprises the town of Crossakeel.

KILTEEL.—Is a parish of County Kildare, Leinster, 5¼ miles East Northeast of Naas. Population, 797.

KILTEEVOCK.—Is a parish of County Donegal, Ulster, 5¼ miles West Northwest of Stranorlar. Area, 41,131 acres. Population, 4,864.

KILTOGHART.—Is a parish of County Leitrim, Connaught, 1¼ miles East Southeast of Leitrim. Area, 30,494 acres. Population, 17,581. It includes part of the town of Carrick-on-Shannon.

KILTONANLEA.—Is a parish of County Clare, Munster, 3¼ miles West Southwest of O'Brien's Bridge. Area, 7,627 acres. Population, 4,016.

KILTOOM.—Is a parish of County Roscommon, Connaught, 5¼ miles Northwest of Athlone. Area, 13,246 acres. Population, 4,150.

KILTUBRID.—Is a parish of County Leitrim, Connaught, 3 miles East Southeast of Drumshambo. Area, 15,608 acres. Population, 7,220.

KILTULLAGH.—Is a parish of County Roscommon, Connaught, 11 miles West Southwest of Castlerea. Area, 24,713 acres. Population, 7,664.

KILVEMNON.—Is a parish of County Tipperary, Munster, 8 miles East Northeast of Fethard. Area, 10,551 acres. Population, 4,983.

KILVOLANE.—Is a parish of County Tipperary, Munster. Area, 8,678 acres. Population, 4,254. It comprises the town of Newport-Tip.

KILWORTH.—Is a market town of County Cork, Munster, on the Funcheon river, 2¼ miles North Northeast of Fermoy. Population, 1,772. It has an ancient Church and Market House, and near it is Moore Park and ruins of Cloughleagh Castle.

KILDARE.—Is an inland County of Leinster. Having County Meath on the North, Counties, Dublin and Wicklow on the East, County Carlow on the South, and Counties, Kings and Queens on the West. Area, 418,436 acres; of which 50,000 are waste. Population in 1851, 96,627. The surface is flat, and the soil is a deep and fertile loam. The Boyne, Barrow and Liffey are the principal rivers. Climate moist, owing to the prevalence of bogs. The Curragh of Kildare, a tract in the centre of this County, is scarcely to be matched for its excellent turf and rich verdure. Wheat, Oats and Barley are the chief crops; farms being generally larger than in other Counties. A large quantity of Peat is sent to Dublin from this County. The County contains 10 baronies and 113 parishes in the dioceses of Kildare and Dublin. Athy, Kildare and Naas are the principal towns. It sends 2 members to the

House of Commons, and has since 1316 given title to the Fitzgeralds, now Dukes of Leinster.

KILDARE.—Is a market town and Episcopal See and Parish of County Kildare, Leinster, on the Great S. and W. Railroad, 30 miles West Southwest of Dublin. Population of parish, 2,654 ; of town, 1,629. It stands on an elevated plain ½ mile from the Curragh of Kildare, It is poorly built and has scarcely any water. It has an ancient Cathedral with Burial Vault of the Fitzgeralds, a part of a Chapel reputed to date from the Fifth Century, a Round Tower, 132 feet in height, the remains of an Abbey and Castle, and a County Infirmery, Roman Catholic Chapel, Nunnery, etc. The celebrated Curragh races take place here in April, June, September and October. The Bishopric is now united with that of Dublin, and comprises 81 parishes in Counties, Kings, Queen and Kildare. It is also the See of a Roman Catholic Bishop.

KILKENNY.—Is an inland County of Leinster, having County Queens on the North, Counties, Carlow and Wexford on the East, and County Waterford on the South and Southwest, and County Tipperary on the West. Area, 508,811 acres, of which 21,000 are waste. Population in 1851, 139,944, mostly all of whom are Romans Catholics. The surface is undulating, gradually sloping toward the South, with several hills which are upward of 1,000 feet in height ; soil varies, but for the most part is light fertile loams. Climate is dryer and crops earlier than the average of Irish Counties. Agriculture is improving, the principal crops being Corn ; but Dairy and Sheep farms are also numerous. Average rent of land 17s. per acre, this being one of the highest rented Counties in Ireland. Manufactures are unimportant. Kilkenny is divided into 9 baronies, and 127 parishes in the diocese of Ossory, Leighlin and Cashel. It sends 3 members to the House of Commons and gives title of Earl to the Butler family.

KILKENNY.—Is a city and municipal and parliamentary borough and capital of County Kilkenny, Leinster, and a County of itself on the Nore river, and on the Irish Southeastern Railroad, 62 miles Southwest of Dublin. Area of city, 921 acres ; of County, 17,012 acres. Population of city, 15,808 ; of borough in 1851, 20,283. It is divided by the river into Irish and English towns, and is, with the exception of its suburb, well built of stone. The streets are paved with block marble. The chief buildings are the Cathedral of St. Canice or Kenny, and of the diocese of Ossory, a cruciform structure of the Fourteenth Century. The Bishop's Palace, Chapter-House, Deanery, fine Round Tower, and Churches of St. John and St. Mary, several Roman Catholic Churches, one of which is the Cathedral of a Roman Catholic Bishop, ruins of a Franciscan Friary, County Court House, County and City Prisons and other public buildings. The Manufactories are much depressed, and the poorer classes are very wretched. Though the city is the residence of many of the provincial gentry, it has some Distilleries, Tanneries, etc. It sends one member to the House of Commons.

KINGS.—Is a County of Leinster. Having County Westmeath on the North, County Kildare on the East, Counties, Queens and Tipperary on the South, and Counties, Tipperary, Galway and Roscommon on the West. Area, 493,985 acres ; of which 145,000 acres are waste. Population in 1851, 112,875. Surface is flat, except in the South, the soil being of the average fertility, being watered by the Shannon, Brosna, Barrow and Boyne rivers. Grazing farms are often extensive, and estates generally large. Average rent of land from 12s. to 15s. per acre. Manufactures are of no importance. The County is divided into 11 baronies and 52 parishes in 5 different dioceses. Birr and Tullamore are the chief towns. This County sends two members to the House of Commons.

KINGSCOURT.—Is a market town of County Cavan, Ulster, 5 miles Southwest of Carrickmacross. Population, 1,614. It has a neat Church and a large Roman Catholic Chapel. In the vicinity is the fine demesne of Cabra Castle.

KINGSTON.—(formerly DUNLEARY.) Is a seaport town and watering place of County Dublin, Leinster, on the bay of Dublin, 7 miles Southeast of Dublin. Population, 10,453. It is finely situated, and the only objects of interest are a short atmospheric railroad to Dalkey, said to be the first ever laid, an Obelisk commemorating the visit of George IV. in 1821, and a fine granite pier, finished at a cost of about £750,000, on which is a revolving light. Latitude, 53° 18′ North. Longitude, 6° 8′ West. Is is a favorite resort of citizens from Dublin.

KINNEGAD.—Is a market town and parish of County Westmeath, Leinster, 12 miles East Southeast of Mullingar. Population of town, 715.

KINNEIGH.—Is a parish of County Cork, Munster. Area, 15,104 acres. Population, 5,530. It comprises the villages of Castletown and Inniskeen.

KINNELTY.—Is a parish of County Kings, Leinster, 4 miles East Northeast of Birr. Area, 13,894 acres. Population, 2,562.

KINSALE.—Is a parliamentary and municipal borough, and seaport town of County Cork, Munster, 13 miles South Southwest of Cork, on the estuary of the Bandon river, called Kinsale harbor. Population, of borough 5,506; including the suburbs of Scilly. The city is built on acclivities and the streets are very steep and narrow. The chief buildings, are a Parish Church of the fourteenth Century, large Roman Catholic Chapel, Convent, Town-Hall, Union Work-House, etc. The fisheries which are valuable form the chief resource of the people. It sends one member to the House of Commons, and gives title of Premier Baron of Ireland to the De'Courcey family, decendants of the Dukes of Normandy, and whose representative has the singular privelege of wearing his hat in the Royal Presence. James II. landed here in 1689. The Old Head of Kinsale is a promontory projecting about 3 miles into the Atlantic, 4½ miles South Southwest, the entrance to Kinsale harbor, and 8 miles South of the town; on it is a Lighthouse with a fixed light, Latitude, 51° 26' 45" North. Longitude, 8° 32' 16" West, at an elevation of 241 feet.

KINRARRA.—Is a small seaport town of County Galway, Connaught, 11 miles South Southeast of Galway. Population, 959.

KIPPURE.—Is a mountain in Counties, Dublin and Wicklow, Leinster, 11 miles South Southwest of Dublin. Elevation, 2,473 feet.

KIRKINRIOLA.—Is a parish of County Antrim, Ulster. Area, 6,390 acres. Population, 8,843. It comprises the town of Ballymena.

KNOCKANE.—Is a parish of County Kerry, Munster, 6 miles West Northwest of Killarney. Area, 57,993 acres. Population, 5,191. It is almost entirely mountainous.

KNOCKANURE.—Is a parish of County Kerry, Munster, 4 miles East Southeast of Listowel. Area, 5,950 acres. Population, 1,358.

KNOCKBREDA.—Is a parish of County Down, Ulster. Area,

8,197 acres. Population, 10,627. It comprises the town of Ballymacarrit.

KNOCKBRIDE.—Is a parish of County Cavan, Ulster, 4½ miles Northeast of Bollieboro'. Area, 18,693 acres. Population, 10,603.

KNOCKGRALTON.—Is a parish of County Tipperary, Munster, 4¼ miles South of Cashel. Area, 9,873 acres. Population, 3,296. It has a singular artificial Mound called the Moat.

KNOCKLADE.—Is a mountain of County Antrim, Ulster, 2¾ miles South of Ballycastle. Height 1,695 feet.

KNOCKMAHON.—Is a village of County Waterford, Munster, near the Atlantic, 1½ miles East Northeast of Bonmahon. Population, 255. It has Copper Mines which in 1840 employed over 1,000 persons and yielded nearly 4,000 tons of Ore.

KNOCKMELEDOWN MOUNTAINS.—Is a range between Counties, Waterford and Tipperary, Munster, and extends East and West for about 18 miles,—their highest point, Knockmeledown, is 4½ miles North Northwest of Lismore, and is 2,690 feet in height.

KNOCKTOPHER.—Is a small market town of County Kilkenny, Leinster, 2 miles East Northeast of Newmarket. Population, 467.

LAGAN.—Is a river of Ulster; it rises in the Slieve-Croob mountains, County Down, and flows Northeast for 35 miles and enters Belfast harbor. It has been made navigable beyond Lisburn, and a canal connects it with Lough Neagh.

LAMBAY.—Is a small fishing island of County Dublin, Leinster, in the Irish sea, 3 miles Southeast of Rush Point. Area, 596 acres. Population, 100.

LANESBOROUGH.—Is a small market town of County Longford, Leinster, on the Shannon river. Population, 300.

LARNE.—Is a market and seaport town of County Antrim, Ulster, on Lough Larne, 17½ miles North Northeast of Belfast. Population, 3,345. It has large manufactories for Cotton, Sail Cloth and Rope, with Bleaching Grounds and Lime Works.

LAVEY.—Is a parish of County Cavan, Ulster, 5 miles Southeast of Cavan. Area, 10,678 acres. Population, 5,931.

LAWRENCE.—(St.) Is a parish of County Limerick, Munster. Area, 280 acres. Population, 1,794. It comprises part of the city of Limerick.

LAYDE.—Is a parish of County Antrim, Ulster, 36 miles North of Belfast. Area, 26,000 acres. Population, 4,208.

LEA.—Is a parish of County Queens, Leinster. Area, 18,188 acres. Population, 7,787. It comprises part of the town of Portarlington. Lea Castle, built in 1260, is here.

LECK.—Is a parish of County Donegal, Ulster, 2 miles East Southeast of Kilkenny. Area, 10,745 acres. Population, 3,122.

LECKPATRICK.—Is a parish of County Tyrone, Ulster. Area, 13,451 acres. Population, 5,723. It comprises part of the town of Strabane.

LEE.—Is a river of County Cork, Munster, rises in Lake Gougane-Barra, flows East and enters Cork harbor after a course of 35 miles. Also a small river of County Kerry, flowing into Tralee Bay.

LEIGHLIN.—(OLD.) Is a decayed village and Episcopal See of County Carlow, Leinster, 2½ miles West of Leighlin Bridge. Population, 140. It has a venerable Cathedral built about 1185. The diocese, founded in 632, is now united to Ossory.

LEIGHLIN BRIDGE.—Is a market town of County Carlow, Leinster, on the Barrow river, 7½ miles Southwest of Carlow. Population, 1,748.

LEINSTER.—Is a province of Ireland. consisting of Counties, Dublin, Kildare, Carlow, Kilkenny, King's and Queen's, Longford, Louth, Meath, Westmeath, Wicklow, and Wexford. Area, 4,876,211 acres. Population, 1,973,731. It gives the title of Duke to the Fitzgerald family, whose head is sole Duke and Premier Peer of Ireland. MOUNT LEINSTER is a mountain between Counties, Carlow and Wexford, Leinster, 5½ miles Southwest of Newtownbarry. Elevation 2,610 feet.

LEITRIM.—Is a maritime County of Connaught. Having Donegal bay on the North, Counties, Fermanagh and Cavan on the East, County Longford on the South, and Counties, Roscommon and Sligo on the West. Area, 392,363 acres; of which 250,000 acres are cultivated, 116,000 acres of mountain and bog land, and nearly 24,000 acres water. Population, 111,808. The surface is mostly wild and rugged, and the soil is poor except in the valleys, where it is often a deep fertile loam. It is watered by the Shannon, Bonnet and Blackwater rivers. Average rent of land, 10s. 8d. per acre. Leitrim contains 5 baronies and 17 parishes, and sends two members to the House of Commons. It gives the title of Earl to the Clements family. Is also the name of a village capital of County Leitrim Connaught, on the Shannon river, 3 miles Northeast of Carrick. Population, 406.

LEIXLIP.—Is a market town and parish of County Kildare, Leinster, at the confluence of the Liffey and Rye rivers, 10 miles West Northwest of Dublin. Population of parish, 2,033 ; of town, 1,086.

LEMANAGHAN.—Is a parish of County Kings, Leinster, 3½ miles East Northeast of Ferbane. Area, 19,615 acres. Population, 5,806.

LETTERKENNY.—Is a market town of County Donegal, Ulster, on the Swilly river, 6½ miles Northwest of Raphoe. Population, 2,161. It is poorly built and has a Church, Session House, etc.

LIFFEY.—Is a river of Leinster, rises in the mountains of County Wicklow about 12 miles Southwest of Dublin, and flows into Dublin bay after a course of 50 miles.

LIFFORD.—Is a market town and capital of County Donegal, Ulster, on the Foyle river, 14 miles South Southwest of Londonderry. The river is here crossed by a Bridge. Population, 752.

LIMERICK.—Is an inland County of Munster ; having County Clare on the North, being separated from it by the Shannon river, Counties, Tipperary and Cork on the East, and County Kerry on the West. Area, with city, 678,224 acres, of which 120,000 acres are waste. Population in 1851, 201.619. Surface is generally mountainous and the soil boggy, except in the center and North, where it is fertile, and the surface flat. It is well watered by the Shannon, Mulkern, Maig and Deel rivers. Dairy and Stock farms are numerous and often large. Limerick is divided into 9 baronies and 125 parishes, and in Dioceses of Limerick, Emly, Killaloe and Cashel. It sends four members to the House of Commons.

LIMERICK.—Is the principal city of West Ireland, and a parliamentary and municipal borough and County of itself, and capital of County Limerick, Munster. It is situated on both sides of the Shannon river, and on an island in that river, 25 miles North Northwest of Tipperary, with which town it communicates by Railroad. Area of borough, 70,000 acres. Population in 1851, 55,268 ; of town, 53,274. It stands on a cultivated plain and is divided into—Englishtown, an old and wretched portion on the island, Irishtown on the South, Newtown on the West and Thomond Gate on the North. Irishtown and Newtown are handsome. The principal buildings are, the Chathedral, the Episcopal Palace, Roman Catholic Cathedral, Schools, County and City Infirmary, Lunatic and Blind Asylums, Fever and other Hospitals ; County and City Court-Houses and Jails, Union Workhouse, etc., and numerous other handsome buildings, including an Exchange, Custom House, etc. The manufactures are very limited, those of Lace and Fish-Hooks being the principal ones. The Corporation consists of a Mayor, (who is Admiral of the Port) Aldermen and Councillors. The Bishopric comprises 88 parishes. The city sends 2 members to the House of Commons, and gives the title of Earl to the Pery family, whose mansion is in the city.

LISBURN.—Is a parliamentary borough, town and parish of Counties, Down and Antrim, Ulster, on the Lagan river, 6½ miles South Southwest of Belfast. Population of parish, 15,015 ; of town, 6,090. It is one of the most beautifully situated and well built towns in Ireland, has a spacious Cathedral, Church, Court House, Fever Hospital, Union Work-House, Linen Hall, etc., with thriving Manufactories of Linen. Markets, Tuesday. The borough sends one member to the House of Commons.

LISCANOR.—Is a village of County Clare, Munster, 4½ miles West of Ennistymon. It is a coast-guard station.

LISNADILL.—Is a parish of County Armagh, Ulster, 2¼ miles Southeast of Armagh. Area, 1,855 acres. Population, 6,282.

LISSAN.—Is a parish of Counties, Tyrone and Londonderry, Ulster, 2 miles Southwest of Moneymore. Area, 24,682 acres. Population, 6,282.

LISSONUFFY.—Is a parish of County Roscommon, Connaught, 44 miles Southeast of Strokestown. Area, 11,665 acres. Population, 4,832.

LISTOWEL.—Is a market town of County Kerry, Munster, on the Feale river, 16¼ miles North Northeast of Tralee. Population, 2,598. It contains a Church, a Bridewell and ruins of a stately Castle, formerly belonging to the Earls of Desmond. It gives the title of Earl to the Hare family.

LITTERMORE.—(or LETTERMORE.) Is an island off the West coast of County Galway, Connaught, on the South side of Kilkinan bay. It has a coast-guard station.

LOGHAR.—Is a parish of County Meath, Leinster, 3½ miles Northwest of Kells. Area, 8,727 acres. Population, 4,495.

LONDONDERRY.—(or DERRY.) Is a maritime County of

Ulster; having the Atlantic Ocean and Loch Foyle on the North, County Antrim on the East, County Tyrone on the South and County Donegal on the West. Area, 518,270 acres; of which 180,000 acres are waste. Population in 1851, 191,744. Surface in the South and centre mountainous, elsewhere lowland; soil mostly fertile; climate mild. Estates are mostly large and owned by the twelve London companies to which the County was granted by James I. The Linen manufacture is extensive. The County is divided into 6 baronies and 31 parishes. It sends four members to the House of Commons.

LONDONDERRY.—Is a fortified city and parliamentary and municipal borough and capital of County Londonderry, Ulster, on the Foyle river; here crossed by a Metal Bridge 1,100 feet in length, 120 miles North Northwest of Dublin. Population in 1851, 19,604. It stands on a ridge projecting into the river and is inclosed by walls and Bastions built in 1609. It has four main streets, which are broad, clean, well paved and lighted. The principal buildings are the Cathedral, Bishop's Palace, a Public and Diocesain Library, a Town Hall, a large Court House, County Jail, Infirmary, Custom House, etc., numerous Industrial and Charitable Schools, Newspapers, Banks, Flour Mills, Distilleries and other manufactories. Markets, Wednesday, Thursday and Saturday. Vessels of 600 tons can ascend the river to the city. Londonderry sends one member to the House of Commons.

LONGFORD.—Is an inland County of Leinster; having Counties, Leitrim and Cavan on the North, County Westmeath on the East and South and County Roscommon on the West. Area, 263,645 acres; of which 58,000 are waste. Population, 83,198. Surface is diversified; soil mostly a rich loam. This County is watered by the Shannon river and Loch Gonna and other Lakes, and is crossed by the Royal Canal. The County is divided into 6 baronies and 23 parishes. It sends two members to the House of Commons.

LONGFORD.—Is a market town and capital of County Longford, Leinster, on the Camlin river, 4 miles from its confluence with the Shannon, and 68 miles West Northwest of Dublin. Population, 4,966. It is well built, clean and thriving, has a handsome Church, Roman Catholic Cathedral, large County Court House, County Jail, Infirmary, Union Work-House, Market House, and large Markets for Grain, Butter and Leather.

LONG ISLAND.—Is an islet off the Southwest coast of County Cork, Munster, in Rowing-water Bay, 6 miles North Northwest of Cape Clear. Length, 2 miles. It is a coast-guard station.

LONGWOOD.—Is a parish of County Meath, Leinster, 9 miles South Southwest of Trim. Population, 587.

LOOP HEAD.—Is a promontory of County Clare, Munster, on the North side of the entrance of the Shannon river. Elevation, 232 feet. It has a Light-House with a fixed light. Latitude, 52° 33′ 39″ North. Longitude, 9° 56′ West.

LORRHA.—Is a parish of County Tipperary, Munster, 5 miles East of Portumna. Area, 16,520 acres. Population, 4,742.

LOUGH-BRICKLAND.—Is a market town of County Down, Ulster, 10 miles North Northeast of Newry. Population, 647. It has the ruins of a Monastery.

LOUGHGALL.—Is a parish of County Armagh, Ulster, 5 miles North Northeast of Armagh. Area, 10,900 acres. Population, 9,615. Castle Dillon is in this parish.

LOUGHGILLY.—Is a parish of County Armagh, Ulster, 3½ miles South Southeast of Markethill. Area, 16,029 acres. Population, 9,852.

LOUGHGLYNN.—Is a parish of County Roscommon, Connaught, 4 miles Northwest of Castlerea. Population, 10,124.

LOUGHGUILE.—Is a parish of County Antrim, Ulster, 8 miles East Southeast of Balleymoney. Area, 29,839 acres; a large portion of which is bog. Population, 6,082.

LOUGHINISLAND.—Is a parish of County Down, Ulster 4¾ miles West Southwest of Downpatrick. Area, 12,485 acres. Population, 6,551.

LOUGHREA.—Is a market town of County Galway, Connaught, 20 miles East Southeast of Galway, on a Lake of the same name. Population, 5,485. It has a neat Parish Church, several Roman Catholic Chapels and Schools, a handsome Nunnery, and a Priory, adjoining the remains of a small Abbey, founded by Richard de Burgh about 1300; also some manufactories of Linen.

LOUTH.—Is a small County of Leinster; having County Armagh and Carlingford Bay on the North, the Irish Sea on the East, County Meath on the South, and Counties, Meath and Monaghan on the West. Area, 206,261 acres; of which 15,500 acres are waste. Population, 91,045; including Drogheda, 107,657. Surface generally fertile; soil good. This County is divided into

4 baronies and 61 parishes in the diocese of Armagh. Drogheda, Dundalk and Ardee, are the principal towns. Louth sends four members to the House of Commons. Is also the name of a decayed town in County Louth, 5½ miles Southwest of Dundalk. Population, 781.

LUGNAGUILLA.—Is a mountain of County Wicklow, Leinster, 6 miles Southeast of Donard. Height, 3,089 feet.

LURGAN.—Is a market town of County Armagh, Ulster, 15 miles East Northeast of Armagh. Population, 4,677. It has handsome Churches and other buildings; and gives title of Baron to the Brownlow family. Markets, Friday. Is also the name of a parish of County Cavan, Ulster. Area, 11,327 acres. Population, 6,557. It comprises the town of Virginia.

LUSK.—Is a parish of County Dublin, Leinster, 3 miles West of Rush. Area, 16,183 acres. Population, 5,961; of village, 872.

MACGILLICUDDY REEKS.—The loftiest mountain range in Ireland—is in County Kerry, Munster, between Loughs, Killarney on the East and Carra on the West, extending for 10 miles. Carrantual is the highest peak, being 3,404 feet high.

MACNEAN.—(UPPER and LOWER.) Are two Loughs of Counties, Fermanagh and Leitrim, Ulster and Connaught, 9 miles Southwest of Enniskillen.

MACOSQUIN.—Is a parish of County Londonderry, Ulster, 3 miles Southwest of Coleraine. Area, 17,804 acres. Population, 6,545.

MACROOM.—Is a market town of County Cork, Munster, 20 miles West of Cork, on the Sullane river. Population, 4,794. It is well situated and has some fine buildings, but consists mostly of Cabins.

MAGHERA.—Is a market town and parish of County Londonderry, Ulster, 20 miles South of Coleraine. Area of parish, 24,792 acres. Population, 14,511; of town, 1,123.

MAGHERACLOONEY.—Is a parish of County Monaghan, Ulster, 4 miles Southwest of Carrickmacross. Area, 14,951 acres. Population, 9,012.

MAGHERACROSS.—Is a parish of Counties, Tyrone and Fermanagh, 5 miles North Northeast of Enniskillen. Area about 10,000 acres. Population, 5,203.

MAGHERACULMONEY.—Is a parish of County Fermanagh,

Ulster. Area, 18,576 acres. Population, 7,021. It has a ruined Abbey and Castle.

MAGHERADROLL.—Is a parish of County Down, Ulster. Area, 12,553 acres. Population, 7,061. It comprises the town of Ballinahinch.

MAGHERAFELT.—Is a market town and parish of County Londonderry, Ulster, 26 miles South of Coleraine. Area of parish, 8,290 acres. Population, 7,649 ; of town, 1,560. It has large Linen manufactories.

MAGHERALIN.—(or MARALIN.) Is a parish of Counties, Down and Armagh, Ulster, on the Laggan river, 3½ miles East Northeast of Lurgan. Area, 8,293 acres. Population, 5,476 ; extensively employed in Linen manufactures and Bleaching.

MAGLASS.—Is a parish of County Wexford, Leinster, 5½ miles Southwest of Wexford. Area, ?,528 acres. Population, 1,112.

MAGUIRE'S BRIDGE.—Is a small market town of County Fermanagh, Ulster, on Colebrook river, 7 miles Southeast of Enniskillen. Population, 685.

MAIN.—Is a river of County Antrim, Ulster, and flows into Lough Neagh after a Southern course of 30 miles.

MAINE.—Is a river of County Kerry, Munster, rises near Castle-Island, and flows West Southwest into Castlemaine harbor. Length, 18 miles.

MALAHIDE.—Is a village and parish of County Dublin, Leinster, 9 miles North Northeast of Dublin. Population, 1,337 ; partly employed in valuable oyster fishing. The village, on a bay of the Irish Sea, is much frequented by bathers.

MALIN HEAD.—Is a promontory of County Donegal, Ulster. Latitude, 55° 22' North. Longitude, 7° 24' West ; with a Signal Tower on its summit.

MANISTER.—Is a parish of County Limerick, Munster, 3 miles East of Croom. Population, 2,946. It has the remains of an Abbey of the twelfth Century.

MANOR-HAMILTON.—Is a market town of County Leitrim, Connaught, 12 miles East of Sligo. Population, 1,507. It is well situated. There is a neat Church and the remains of a fine Baronial Castle here.

MARKETHILL.—Is a market town of County Armagh, Ulster, 6 miles Southeast of Armagh. Population, 1,424. It has a neat

Court House, Jail, and a mansion of Lord Gosford, who owns the town.

MARK.—(ST.) Is a parish of County Dublin, Leinster. Area, 351 acres. Population, 15,234. It comprises part of the city of Dublin.

MARYBOROUGH.—Is a borough and town of County Queens, Leinster, on an affluent of the Barrow river, 10 miles South Southwest of Portarlington, and 53¾ miles Southwest of Dublin. Population, 3,633. Markets, Thursday.

MAYNOOTH.—Is a market town of County Kildare, Leinster, on the Royal Canal, 15 miles West Northwest of Dublin. Population, 2,201. It has the remains of a Castle (formerly chief seat of the Fitzgeralds,) a large Roman Catholic Chapel and Convent, and the Royal College of St. Patrick, founded in 1795 for the education of the Roman Catholic Clergy, which accommodates 450 students—250 of whom are maintained free. Carton, the demesne of the Duke of Leinster, is in this vicinity.

MAYO.—Is a maritime County of Connaught; having Counties, Sligo and Roscommon on the East, County Galway on the South, and the Atlantic on the West and North. Area, 1,363,882 acres, of which 800,000 acres are waste. Population, 274,716. Surface is mostly mountainous but comprises many fertile and level tracts; soil mostly light and with the moist climate is better suited for grazing than for tillage; it is watered by the Moy river. Estates large, and farms very small, the tenants of which are very poor. The Fisheries are very valuable. The coast line is fringed with cliffs and islets and indented with innumerable bays, the largest of which are Killala, Broadhaven, Blacksod and Clew bays and Killery harbor. The manufactories are small. The County is divided into 9 baronies and 60 parishes in the dioceses of Tuam, Killala, Achonry and Elphin. Chief towns, Castlebar, Ballina and Westport. It sends two members to the House of Commons. Is also the name of a parish in County Mayo, 3 miles Southeast of Ballagh. Area, 11,848 acres. Population, 4,179.

MEATH.—Is a maritime County of Leinster; having County Dublin and the Irish sea on the East, and Counties, Louth, Monaghan, Kings, Cavan, Kildare and Westmeath on its other sides. Area, 579,899 acres; nearly the whole of which is cultivated. Population in 1851, 140,750. Surface slightly undulating, and soil mostly a rich clay loam; watered by the Boyne river. Tillage farms cover more than four-fifths of the County, and the condition of the numerous small farmers is wretched. The County is divided into 12 baronies and 147 parishes in the diocese of Meath, founded about 1150. The chief towns are Trim, Kells and Naoan. This County sends two members to the House of Commons. Previous to the Anglo-Norman conquest the King of Meath was supreme Monarch of Ireland.

MEELICK.—Is a parish of County Galway, Connaught, 2 miles Southeast of Eyrecourt. Area, 4,292 acres. Population, 1,710. Is also the name of a parish in County Mayo, Connaught, 3 miles West Southwest of Swinford. Area, 8,062 acres. Population, 3,915.

MEVAGH.—Is a parish of County Donegal, Ulster, 7 miles Northwest of Millford. Area, 21,026 acres. Population, 5,620.

MIDDLETON.—Is a market town of County Cork, Munster, 14 miles East of Cork, on a stream of same name. Population, 4,591. It is neatly built, and its port about 1 mile nearer Cork harbor is reached by vessels of 200 tons. It gives the title of Viscount to the Broderic family. Is also the name of a village of County Armagh, Ulster, 7 miles Southwest of Armagh. Population, 708.

MILLIFONT.—(or MELLEFONT.) Is a parish of County Louth, Leinster, 5 miles West Northwest of Drogheda, where are the remains of a celebrated Abbey.

MILLTOWN.—There are two villages by this name, as follows, viz: County Kerry, Munster, 8 miles South of Tralee. Population, 787. (Kilcoleman Abbey is in this vicinity) County Dublin, Leinster, 2 miles South Southeast of Dublin. Population, 726.

MILLTOWN-MALBAY.—Is a town of County Clare, Munster, on the cove of Malbay, 8 miles West of Ennis. Population, 1,295.

MITCHELLSTOWN.—Is a market town of County Cork, Munster, 25 miles North Northeast of Cork. It is finely situated. It has a large Square, handsome Church and a College for indigent persons of superior station adjoining seat of the Earl of Kingston. Is also the name of a parish of County Meath, Leinster, 3 miles Southeast of Nobber. Area, 973 acres. Population, 248.

MITCHELLSTOWN CAVES.—Is in County Tipperary, Munster, 7 miles East Northeast of Mitchellstown. They are a large and fine series of stalactitic caverns.

MOATE.—Is a market town of County Westmeath, Leinster, on the Grand Canal, 9 miles East Southeast of Athlone. Population, 2,095.

MODREENY.—Is a parish of County Tipperary, Munster. Area, 12,165 acres. Population, 5,286. It comprises the town of Cloughjordan.

MONAGAY.—Is a parish of County Limerick, Munster. Area, 22,701 acres. Population, 6,366. It comprises part of the town of Newcastle.

MONAGHAN.—Is an inland County of Ulster; having County Tyrone on the North, County Armagh on the East, Counties, Louth and Meath on the South, and Counties, Cavan and Fermanagh on the West. Area, 327,078 acres; of which 20,000 acres are waste. Population in 1851, 143,418. Surface is hilly, interspered with many bogs and small lakes; climate moist. Chief crops are Flax, Oats, Wheat and Potatoes, but of an inferior quality. The Blackwater is the principal river of this County. The manufacture of Linen, formerly very flourishing, has greatly declined. The Ulster Canal traverses the County, which is divided into 5 baronies and 19 parishes in the diocese of Clogher. Monaghan, Clones and Carrickmacross are the chief towns. This County sends two members to the House of Commons.

MONAGHAN.—Is a market town and capital of the above County, on the Ulster Canal, 68 miles North Northwest of Dublin. Area of parish, 13,547 acres. Population, 12,160; of town, 4,130. It is situated on the borders of two large ponds. It contains some good buildings, Churches, etc. Linen and Hog Markets on Monday.

MOIRA.—Is a small market town of County Down, Ulster, 14 miles Southwest of Belfast. Population, 823; chiefly employed in the manufacture of Linen. It gives title of Earl to the Marquis of Hastings,

MONALTY.—Is a parish of County Meath, Leinster, on a small river of same name, 14 miles North Northwest of Navan. Area, 12,678 acres. Population, 6,279.

MONASTEREVEN.—Is a market town of County Kildare, Leinster, 6 miles West Southwest of Kildare, on the Barrow river, and a branch of the Grand Canal. Population, 1,099. It has an elegant Church and several Docks and Storehouses.

MONEY-GALL.—Is a market-town of County Kings, Leinster, 8 miles Southwest of Roscrea. Population, 764.

MONEYMORE.—Is a market town of County Londonderry, Ulster, 30 miles South of Coleraine. Population, 942.

MONIVÆ.—Is a parish of County Galway, Connaught, 5 miles North Northeast of Athenry. Area, 21,932 acres. Population, 4,810. It has several ruined Churches and Castles.

MONKSTOWN.—There are several parishes by this name, as follows, viz: County Cork, Munster, on Cork harbor, 8 miles East Southeast of Cork. Area, 1,541 acres. Population, 2,138 ; County Dublin, Leinster, comprising Kingstown ; (½ mile West of which is the hamlet Monkstown,) also several villages and numerous residences. Population of parish, 13,143 ; County Meath, Leinster, 5 miles East Southeast of Navan. Area, 1,870 acres. Population, 460.

MONKSLAND.—Is in County Waterford, Munster. Area, 2,118 acres. Population, 1,672. It comprises the village of Knockmahon.

MOORE.—Is a parish of County Roscommon, Connaught, 4¼ miles East Northeast of Ballinasloe. Area, 21,013 acres. Population, 4,608.

MORNINGTON.—Is a hamlet of County Meath, Leinster, on the Boyne river, 2¾ miles East Northeast of Drogheda. Population, 188. It gives the title of Earl to the Wellsley-Pole family.

MOTHELL.—Is a parish of County Waterford, Munster, 2¼ miles Southeast of Carrickbeg. Area, 20,740 acres. Population, 3,723. It has the remains of a Castle and ancient Abbey.

MOUNT-LEINSTER.—Is a mountain of County Carlow, Leinster, 7 miles East Northeast of Borris. Height, 2,610 feet.

MOUNT-MELLICK.—Is a market town of County Queens, Leinster, on a branch of the Grand Canal, 6 miles Northwest of Maryborough. Population, 4,755 ; mostly employed in Cotton and Woolen Manufactories, Iron and Brass Works and Potteries. It has a branch Bank. There are two Markets weekly.

MOUNTRATH.—Is a market town of County Queens, Leinster, 14 miles East Northeast of Roscrea. Population, 3,000. It has a Monastery and Nunnery, several Schools, etc., also Manufactories of Cotton and Worsteds.

MOURNE ABBEY.—Is a parish of County Cork, Munster, 5 miles South Southeast of Mallow. Area, 11,436 acres. Population, 4,154. It has the ruins of a Preceptory of the Knights Templar.

MOURNE.—Are a range of mountains in County Down, Ulster, extending 14 miles from East to West, between Newcastle on the Irish Sea and Carlingford bay. Their highest summits rise to between 2,000 and 3,000 feet above the sea.

MOURNE.—Is a river of County Donegal, Ulster, joining the Foyle river at Lifford after a course of 8 miles.

MOVILLE.—Is a small market town and parish of County Donegal, Ulster, on Lough Foyle, 18 miles North Northeast of Londonderry. Area of parish, 15,950. Population, 6,016; of town, 595. (UPPER.) Is also the name of a parish adjoining the above on the South. Area, 19,031 acres. Population, 5,069.

MOY.—Is a river rising in County Sligo, Connaught, and flows North and West through Counties, Sligo and Mayo, and after a course of about 40 miles enters Killala bay. It is navigable to near Ballina.

MOY.—Is a market town of County Tyrone, Ulster, 6 miles West Northwest of Armagh. Population, 757. It has Episcopal, Presbyterian and Independent Churches, and Methodist and Roman Catholic Chapels. There are horse and cattle Fairs held here the first Friday of each month. The Moy valley station of the Midland and Great Western Railroad is 5½ miles East of Kinnegad and 30½ miles West of Dublin.

MOYACOMBER.—Is a parish of Counties, Carlow, Wexford and Wicklow, Leinster. Area, 17,434 acres. Population, 4,933. It comprises the village of Clonegal.

MOYARTA.—(or MOYFESTA.) Is a parish of County Clare, Munster, on the peninsular between the Atlantic and the estuary of the Shannon river, 10 miles East Northeast of Loop-Head. Area, 15,613 acres. Population, 8,697.

MOYCULLEN.—Is a parish of County Galway, Connaught, 6½ miles Northwest of Galway. Area, 35,824 acres. Population, 6,420.

MOYLOUGH.—Is a parish of County Galway, Connaught, 12 miles East of Tuam. Area, 23,386 acres. Population, 7,248.

MOYRUS.—Is a parish of County Galway, Connaught. Area, 101,510 acres; including lakes and mountains. Population, 11,969.

MUCKISH.—Is a mountain of County Donegal, Ulster, 5 miles South of Dunfanaghy. Height, 2,190 feet.

MUCKNO.—Is a parish of County Monaghan, Ulster. Area,

17,104 acres ; including Lough Muckno. Population, 9,902. It comprises the town of Castle-Blayney.

MUCKROSS.—Is a peninsular of County Kerry, Munster, between the middle and lower Loughs of Killarney, with the ruins of an Abbey founded in 1,440.

MUFF.—Is a parish of County Donegal, Ulster, on Lough Foyle, 6 miles North Northeast of Londonderry. Area, 15,030 acres. Population, 4,037.

MUILREA.—(or MULREA.) Is a mountain range of County Mayo, Connaught, extending along the North of Killery harbor. Elevation of highest peak, 2,688 feet.

MULLAGH.—Is a parish of County Cavan, Ulster, 7 miles East Southeast of Virginia. Area, 12,872 acres. Population, 6,526.

MULLAGHBRACK.—Is a parish of County Armagh, Ulster. Area, 11,557 acres. Population, 8,570. It comprises part of the town of Markethill. The inhabitants are extensively employed in the manufacture of Linen.

MULLAGHMORE.—Is a promontory of County Sligo, Connaught, 13 miles North of Sligo, projecting North into Donegal bay. Its proprietor, Lord Palmerston has built a harbor and small fishing village on its East side.

MULLET.—Is a peninsular of the West coast of County Mayo, Connaught, connected with the mainland by a narrow isthmus from which it extends North and South at right angles.

MULLINGAR.—Is a market town and parish of County Westmeath, Leinster, and capital of said County, on the Brosna river and the Royal canal, 50 miles West Northwest of Dublin. Area of parish, 22,322 acres. Population, 9,903 ; of town, 5,516. It is well built, except in the suburbs, and has a handsome parish Church, Roman Catholic Cathedral, a Convent, Schools, Market House, etc. Markets, Thursday. There are two Fairs held here Annually.

MULLIN'S.—(ST.) Is a parish of Counties, Carlow and Waterford, Leinster and Munster, 9 miles South of Burris. Area, 21,202 acres. Population, 6,769. It comprises the villages of Tinnahinch and Ballymurphy.

MULROY BAY.—Is a deep inlet of the Atlantic Ocean, County Donegal, Ulster, between Sheephaven bay and Lough Swilley. Length, 12 miles. Its shores are remarkably beautiful.

MULTIFARNHAM.—Is a parish of County Westmeath, Leinster, 6 miles North Northwest of Mullingar. Area, 4,895 acres. Population, 1,366.

MUNSTER.—Is the most Southern and largest of the four provinces of Ireland. Area, 6,064,579 acres. Population, 2,396,161. Surface is highly diversified, and the soil is watered by the Shannon river on the North, and the Suir river on the East. This province is divided into the Counties, Clare, Kerry, Limerick, Cork, Tipperary and Waterford. Before the Norman conquest it was separated into the Kingdoms of North and South Munster.

MUSKERRY.—Is a mountainous district of County Cork, Munster. Area about 311,000. Population, 90,511.

NAAS.—Is a market town and parish of County Kildare, Leinster, on a branch of the Grand Canal, 9 miles Southwest of Dublin. Area of parish, 5,526. Population, 4,863; of town, 3,471. It gives title of Viscount to the Earl of Mayo.

NANTINAN.—(or NANTENANT.) Is a parish of County Limerick, Munster, 2½ miles South Southeast of Askeaton. Area, 7,922 acres. Population, 3,018.

NARRAGHMORE.—Is a parish of County Kildare, Leinster, 2 miles North Northwest of Ballytore. Area, 11,270 acres. Population, 5,895.

NAVAN.—Is a market town and parish of County Meath, Leinster, at the confluence of the Boyne and Blackwater rivers, 26 miles Northwest of Dublin. . Area of parish, 3,544 acres. Population, 6,834; of town, 4,987. It has a Court House, Mills, etc.

NEAGH.—(LOUGH.) Is a lake on the border of County Antrim, Ulster. Area, 98,255 acres. Length, 17 miles. Breadth, 10 miles. Mean depth, 40 feet. It receives the Upper Bann and Blackwater rivers from the South, and discharges its surplus waters on the North by the Lower Bann. Height, 48 feet above the sea at low water.

NENAGH.—Is a market town and parish of County Tipperary, Munster, near the Nenagh river, an affluent of the Shannon river, 13 miles Northeast of Newport. Area of parish, 3,881. Population, 9,540; of town, 7,235. It is situated in a district of great beauty and fertility, and is neat, clean and thriving.

NEWBLISS.—Is a thriving market town of County Monaghan, Ulster, 4 miles East Southeast of Clones. Population, 566.

NEWBRIDGE.—Is a market town of County Kildare, Leinster,

2 miles East Northeast of Kildare on the Liffey river. Population, 792. Near it are the ruins of Great Connell Abbey.

NEWCASTLE.—There are two towns and several parishes by this name. One of the towns is in County Limerick, Munster, 25 miles Southwest of Limerick. Population, 2,917. It has a neat Church and other buildings ; another is a seaport town of County Down, Ulster, 11 miles South Southwest of Downpatrick. Population, 1,157. It is frequented as a watering place. The parishes are situated as follows, viz : County Limerick, Munster. Area, 5,325 acres. Population, 4,191. It comprises part of the town of Newcastle ; County Tipperary, Munster, 7 miles Southwest of Clonmel. Area, 10,855 acres. Population, 2,293 ; of village 253 ; (UPPER.) Is in County Wicklow, Leinster. Area, 7,026 acres. Population, 2,766. It comprises the town of Newtown-Mount-Kennedy ; County Waterford, Munster, 4 miles Northeast Kilmacthomas. Area, 3,961 acres. Population, 1,337. (LOWER.) Is in County Wicklow, Leinster, 2½ miles Southeast of Newtown-Mount-Kennedy. Area, 4,750 acres. Population, 1,226 ; of village 196. (NEWCASTLE-LYONS.) Is in County Dublin, Leinster, 10 miles West Southwest of Dublin. Area, 4,282 acres. Population, 1,108.

NEWMARKET.—Is a market town of County Clare, Munster, 4½ miles South Southeast of Clare, on the Fergus river. Population, 1,526.

NEWPORT-PRATT.—Is a seaport town of County Mayo, Connaught, on the Newport river, 8 miles West Northwest of Castlebar. Population, 1,091. The harbor is spacious and safe.

NEWPORT-TIP.—Is a market town of County Tipperary, Munster, on the Mulkern river, 9½ miles Northeast of Limerick. Population, 1,072.

NEWRY.—Is a parliamentary borough, river-port town and parish of Counties, Down and Armagh, Ulster, on the Newry Water, here crossed by 8 bridges 6 miles above its fall in Carlingford bay, and on the Newry Canal. Area of parish, 5,470 acres. Population, 25,168. Area of borough, 2,543 acres. Population, 13,227 ; of town, 13,556. It is well built, except some parts of the old town. It contains many handsome Churches and other buildings. The borough sends one member to the House of Commons.

NEWTOWN.—There are numerous parishes, villages and a small town by this name, the principal of which is a parish of County

Meath, Leinster, 3 miles North of Kells; another is in County Westmeath, Leinster. Area, 10,249 acres. Population, 3,010. It comprises the town of Tyrrell's Pass.

NEWTON-ARDES.—Is a borough, seaport town and parish of County Down, Ulster, at the Northern extremity of Lough Strangford, 10 miles East of Belfast. Area of parish, 14,804 acres. Population, 13,886; of town, 9,551. It is well built and has good buildings. The weaving and embroidering of Damask Muslins is carried on here to a considerable extent.

NEWTOWN-BARRY.—Is a market town and parish of County Wexford, Leinster, at the confluence of the Clady with the Slaney river 3 miles South of Clonegal. Area of parish, 8,284. Population, 3,723; of town, 1,437.

NEWTOWN-CLONEBURN.—Is a parish of County Meath, Leinster, on the Boyne river, 1 mile East of Trim. Area, 566 acres. Population, 298.

NEWTOWN-CROMMOLIN—Is a parish of County Antrim, Ulster, 3 miles Northeast of Clough. Area, 3,466 acres. Population, 799; of village, 175.

NEWTOWN-FORBES.—Is a parish of County Longford, Leinster, 2½ miles West Northwest of Longford. Population, 478.

NEWTOWN-HAMILTON.—Is a market town and parish of County Armagh, Ulster, 9½ miles Southeast of Armagh. Area of parish, 12,405 acres. Population, 7,538; of town, 1,231.

NEWTOWN-LENANT.—Is a parish of County Tipperary, Munster, 3 miles Northeast of Carrick-on-Suir. Area, 5,774 acres. Population, 1,806.

NEWTOWN-MOUNT-KENNEDY.—Is a market town of County Wicklow, Leinster, 8 miles North Northwest of Wicklow. Population, 823.

NEWTOWN-LIMAVADDY.—Is a disfranchised borough and market town of County Londonderry, Ulster, 15 miles Northeast of Londonderry, on the Roe river. Population, 3,101. It is well built and has many fine public buildings. Markets, Monday, Tuesday and Friday.

NEWTOWN-STEWART.—(formerly LISLAS.) Is a market town of County Londonderry, Ulster, on the Mourne river, 5 miles North of Gorton. Population, 1,405. Fairs, last Monday of every Month. About 1½ miles Southwest of the town is Barons Court, the seat of the Marquis of Abercorn.

NOHOVALL.—(or NOUGHILL.) There are numerous parishes by this name, as follows, viz : Counties, Westmeath and Longford, Leinster, about 4 miles West Northwest of Ballymore. Area, 15,152 acres. Population, 4,480 ; (N.-DALY,) Counties, Cork and Kerry, Munster, 6½ miles West Northwest of Mill Street. Area, 17,373 acres. Population, 3,954 ; County Cork, Munster, 4 miles East of Kinsale. Area, 2,568 acres. Population, 1,175 ; of village, 142 ; (N.-KERRY,) County Kerry, Munster, 2½ miles West of Castle-Island. Area, 3,204 acres. Population, 944 ; County Clare, Munster, 2 miles Northeast of Kiltinora. Area, 4,661 acres. Population, 450.

NURNEY.—There are several parishes by this name in Leinster, as follows, viz : County Carlow, 3 miles Northeast of Leighlin-Bridge. Area, 2,723 acres. Population, 905 ; County Kildare, 4 miles Southwest of Kildare. Area, 1,798 acres. Population, 735 ; 3 miles North Northwest of Carbery. Area, 2,130 acres. Population, 651.

O'BRIEN'S BRIDGE.—Is a parish of County Clare, Munster, 4 miles South Southwest of Kilaloe. Area, 11,425 acres. Population, 4,995 ; of whom 435 are in the village, which is on the Shannon river.

O'DORNEY.—Is a parish of County Kerry, Munster, 4 miles North of Tralee. Area, 7,227 acres. Population, 3,142.

OFFERLANE.—Is a parish of County Queens, Leinster, 3½ miles West Southwest of Mountrath. Area, 48,927 acres. Population, 10,491. It has the remains of 3 baronial Castles.

OGONNELLOE.—(or OGONILLOE.) Is a parish of County Clare, Munster, 4 miles Northwest of Killaloe. Area, 9,926 acres. Population, 3,162. It comprises part of Lough Derry.

OLDCASTLE.—(or CLOTYNGE.) Is a market town and parish of County Meath, Leinster, 12 miles West Northwest of Kells. Area of parish, 7,908. Population, 5,079 ; of town, 1,508. It has the largest Yarn trade in the County, also extensive Corn Mills.

OMAGH.—Is a market town of County Tyrone, Ulster, 27 miles South of Londonderry. Population, 2,947. It is well built and clean, and contains an elegant County Court House, Jail, etc. The town was destroyed by fire in 1689 and 1743. Markets, weekly. Fairs, first Tuesday of every month. It is the seat of the Courts of Assize for the County.

OMEY.—(or UMMA.) Is a parish of County Galway, Connaugh'

Area, 20,836 acres. Population, 7,953. It comprises the town of Clifden.

OOLA.—(or ULLA.) Is a parish of County Limerick, Munster, 5 miles Northwest of Tipperary. Area, 6,859 acres. Population 3,377 ; of village, 398.

ORANMORE.—Is a town and parish of County Galway, Connaught, 5 miles East of Galway, at the head of Oranmore bay. Area, 19,335 acres ; including Loughs. Population, 7,952 ; of town, 842. It has a handsome parish Church, Roman Catholic Chapel and a Castle of the fifteenth Century. It has a large traffic in Turf, Sea Manure and Fish.

ORREY.—(or KILMORE.) Is a barony of County Cork, Munster. Area, 69,346 acres. Population, 34,134. It gives the title of Earl to the Boyle family, Earls of Cork.

OSSORY.—Is an old principality and diocese of Counties, Kilkenny, Queens and Kings, Leinster. Since 1833 the Protestant as well as Roman Catholic See has had its seat at Kilkenny.

OUTRAGH.—(or OUTRATH.) There are several parishes by this name, as follows, viz : County Leitrim, Connaught. Area, 21,690 acres. Population, 9,255. It comprises the town of Ballinamore ; County Tipperary, Munster, 3 miles North Northeast of Cahir. Area, 1,548 acres. Population, 554 ; County Kilkenny, Leinster, 2 miles Southeast of Kilkenny. Area, 2,050 acres. Population, 599.

OVOCA.—(or AVOCA.) Is a river of County Wicklow, Leinster, formed by the confluence of the waters Avonbeg and Avonmore, and after a Southeast course of 6 miles enters the Irish Sea near Arklow. Its vale is celebrated for its picturesque beauty.

PALLAS-KENRY.—Is a market town of County Limerick, Munster, 10 miles East Southeast of Limerick. Population, 783. Fair on the fifteenth of August. County Petty Sessions are held here.

PALLICE.—(or PALLAS.) Is a hamlet of County Longford, Leinster, 1½ miles Southeast of Ballymahon. Oliver Goldsmith was born here in 1731.

PALMERSTOWN.—Is a town and parish of County Dublin, Leinster, adjoining Phœnix Park. Area, 1,517 acres. Population, 1,411 ; of village, 201. It gives the title of Viscount to the Temple family.

PARSONSTOWN.—Is a parish of County Louth, Leinster,

4 miles East Southeast of Dunleer. Area, 524 acres. Population, 237.

PASSAGE.—There are two small seaport towns by this name in Munster. One is in County Waterford, 6 miles East Southeast of Waterford, on the estuary of the Suir river. Population, 624. It is irregularly built, and has a Pier and a Block House. Another is in County Cork, Munster, 7½ miles East Southeast of Cork, on the estuary of the Lee river, opposite Great Island. Population, 1,721. It has several Churches. It is the port of Cork for all large shipping, and is the seat of Petty Sessions. There is also a village and hamlet of same name in County Cork.

PIGOEPET.—Is a village of Counties, Donegal and Fermanagh, Ulster, on the Yermon river, near its mouth in Lough Erne, 13 miles Southeast of Donegal. Population, 616. It stands amidst wooded hills and is a station for the numerous pilgrims who resort to Lough Dreg.

PHILIPSTOWN.—Is a market town and Assize town of County King's, Leinster, on the Grand Canal, 9 miles East Northeast of Tullamore. Population, 1,489. The principal edifices are an old Castle, (once the residence of King Philip of Spain, and now a barracks,) two Schools, Session House and Jail. It is also the name of a parish in County Louth, Leinster, 4 miles Northwest of Ardee. Area, 3,660 acres. Population, 1,669.

PHILIPSTOWN.—(or NUGENT.) Is a parish 4 miles West Northwest of Dundalk. Area, 1,036 acres. Population, 401.

PILLTOWN.—Is a small market town of County Kilkenny, Leinster, 4 miles East of Carrick-on-Suir. Population, 701. It is clean and neat, and adjoining it is the seat of the Earl of Besborough.

POMEROY.—Is a village and parish of County Tyrone, Ulster, 9 miles Northwest of Dungannon. Area, 15,950 acres. Population, 8,527; of village 491.

PORTADOWN.—Is a market town of County Armagh, Ulster, 10 miles East Northeast of Armagh on the Bann river. Population, 2.505. Markets, weekly. There are fifteen Fairs held here Annually. There are large manufactories of Linen and Cotton goods, and a large Distillery here. It has a brisk trade in Corn.

PORTAFERRY.—Is a seaport and market town of County Down, Ulster, near the entrance to Lough Strangford, 7½ miles

East Northeast of Downpatrick. Population, 2,007. There are fourteen Fairs held here Annually.

PORTARLINGTON.—Is a parliamentary and municipal borough and town of Counties, Kings and Queens, Leinster, on the Barrow river, 40 miles West Southwest of Dublin. Area of borough, 915 acres. Population, 3,106. It is one of the best built and cleanest country towns in Ireland, and has some fine buildings. It sends one member to the House of Commons. It gives title of Earl to the Dawson family.

PORTGLENONE.—Is a market town in County Antrim, Ulster, on the Bann river, 7 miles South Southeast of Kilrea. Population, 990.

PORTMARNOCK.—Is a parish of County Dublin, Leinster, 8½ miles North Northeast of Dublin. Area, 2,084 acres. Population, 631.

PORTRUSH.—Is a small seaport town of County Antrim, Ulster, 5 miles North of Coleraine, on a narrow peninsular, near the Skerry island. Population, 630. It is neatly built, and is much resorted to for bathing.

PORT-STEWART.—Is a maritime town of County Londonderry, Ulster, 4 miles North Northwest of Coleraine. Population, 603. It is frequented as a watering place.

PORTUMNA.—Is a market town of County Galway, Connaught, 17 miles East Southeast of Lough Rea. Population, 1,643. It has a good trade.

POWERSCOURT.—(or STAGONIL.) Is a parish of County Wicklow, Leinster, 3 miles West Southwest of Bray. Area, 18,938 acres. Population, 3,070. It gives the title of Viscount to the Wingfield family. Near it is a Waterfall.

PUFFIN ISLAND.—Is a rocky island off the coast of County Kerry, Munster, in St. Finnan's bay, 3 miles Southeast of Breahead, Valentia island.

QUEENS.—(COUNTY.) Is an inland County of Leinster; having County Kings on the North, County Kildare on the East, County Kilkenny on the South and County Tipperary on the West. Area, 424,854 acres; of which 69,289 acres are uncultivated. Population, 111,623. The Surface is mostly flat, rising in the Northwest into the Slichbloom mountains. The soil is fertile, interspersed with bog, and watered by the Barrow and Nore rivers. Agriculture has improved. Dairy and other stock are numerous. Estates mostly large. There are some manufactories of Woolen, Cotton and Linen goods. It is divided into 11 baronies and 53 parishes. It sends two members to the House of Commons.

QUEENSTOWN.—(Now called Cove of Cork.)

QUIN.—Is a parish of County Clare, Munster, 5 miles East of Clare. Area, 9,585 acres. Population, 3,634; of village, 173.

QUINCE.—(or Squince.) Is a small island off the Southwest coast of County Cork, Munster, Southwest of entrance to Glandore harbor. It has good pasturage.

RACAVAN.—Is a parish of County Antrim, Ulster. Area, 17,563 acres. Population, 5,356. It comprises the town of Broughshane.

RAHAW.—(or Raghan.) There are two parishes by this name. One is in County Cork, Munster. Area, 10,083 acres. Population, 4,061. It comprises the town of Ballymagooley. The other is in County Kings, Leinster, on the Grand Canal, 5 miles West of Tullamore. Area, 14,985 acres. Population, 4,311. It has numerous ruins.

RAHENY.—(or Ratheny.) Is a village and parish of County Dublin, Leinster, 4 miles Northeast of Dublin, on the North side of Dublin bay. Population of village, 295.

RAHOON.—Is a village and parish of County Galway, Connaught. Area, 15,168 acres. Population, 14,443. It comprises part of the town of Galway.

RAMOAN.—(or Rathmoan.) Is a parish of County Antrim, Ulster, at its Northeast extremity. Area, 12,066 acres. Population, 3,110. It comprises the town of Ballycastle with Kenbane headland and the mountain Knocklayd.

RANDALSTOWN.—Is a market town and formerly parliamentary borough of County Antrim, Ulster, 2½ miles North of Lough Neagh and 5 miles West Northwest of Antrim, on the Main river near its mouth. Population, 588. The town is neat and has a good Market House, with Assembly Rooms, Church, Barracks and some manufactories. Shane's Castle, the seat of Earl O'Neill is adjoining. Linen markets first Wednesday in every month.

RANELAGH.—Is a suburb of Dublin, 1¾ miles South Southeast of Dublin Castle. It is well built and gives the title of Earl to the Jones family.

RAPHOE.—Is an episcopal market town, parish and barony of

County Donegal, Ulster, 5½ miles West Northwest of Lifford. It consists chiefly of a market place and has some Public Buildings, Churches, etc. Market, weekly. The diocese comprises 35 parishes in County Donegal. Raphoe is also the head of the Roman Catholic diocese, and a monastery is said to have been founded here by St. Columb of Iona.

RASHARKIE.—Is a parish of County Antrim, Ulster, 6 miles North of Portglenone. Area, 19,337 acres. Population, 7,507.

RATASS.—(or Rathass.) Is a parish of County Kerry, Munster. Area, 2,365 acres. Population, 2,838. It comprises part of the town of Tralee.

RATH.—Is a parish of County Clare, Munster, 2 miles South Southwest of Corrofin. Area, 8,489 acres. Population, 2,647. It is also the name of a village of County Kings, Leinster, 5 miles Southwest of Frankford.

RATHANGAN.—Is a market town and parish of County Kildare, Leinster, 5 miles North Northwest of Kildare, on the Blackwood river and branch of the Grand Canal. Area, 11,530 acres. Population, 2,991 ; of town, 1,083.

RATHASPECK.—There are several parishes by this name in Leinster, as follows, viz : County Westmeath. Area, 7,664 acres. Population, 2,135. It comprises the town of Rathowen ; County Wexford, 3 miles South Southwest of Wexford. Area, 2,804 acres. Population, 737.

RATHASPECK.—(or Rathasback.) Chiefly in County Queens, 5 miles South Southwest of Athy. Area, 8,218 acres. Population, 4,133.

RATHBOURNEY.—Is a parish of County Clare, Munster, 2½ miles South Southwest of Ballyvaughan. Area, 9,633 acres. Population, 1,000.

RATHBRAN.—Is a parish of County Wicklow, Leinster. It comprises the town of Stratford-on-Slaney.

RATHCAVAN.—Is a parish of County Antrim, Ulster. Area, 17,563 acres. Population, 5,356. It comprises the town of Broughshane.

RATHCLARIN.—Is a parish of County Cork, Munster, 4½ miles Southeast of Bandon. Area, 5,901 acres. Population, 2,907.

RATHCLINE.—Is a parish and barony of County Longford, Leinster. Area of parish, 12,883 acres. Population, 2,792. The parish comprises the town of Lanesboro.

RATHCONNEL.—Is a parish of County Westmeath, Leinster, 3½ miles East Northeast of Mullingar. Area, 15,699 acres. Population, 3,605. Rathconnel bog has an area of 2,505 acres.

RATHCONRATH.—Is a village, parish and barony of County Westmeath, Leinster. The village is situated 8½ miles West of Mullingar. Area of parish, 8,745 acres. Population, 3,378. The residence of the D'Alton family is in this parish, which has many antiquities.

RATHCOOLE.—There are several parishes by this name, as follows, viz: County Kilkenny, Leinster, 3½ miles Northeast of Kilkenny. Area, 3,672 acres. Population, 1,283 ; County Dublin, Leinster, 11 miles West Southwest of Dublin. Area, 4,705 acres. Population, 1,527 ; County Tipperary, Munster, 2 miles Northwest of Fethard. Area, 5,904 acres. Population, 1,677.

RATHCOONAY.—Is a parish of County Cork, Munster, 4 miles East Northeast of Cork. Area, 5,152 acres. Population, 3,376.

RATHCORE.—Is a parish of County Meath, Leinster, on the Royal Canal. Area, 12,804 acres. Population, 3,546. It comprises the town of Enfield.

RATHCORMACK.—Is a market town and parish, and formerly a parliamentary borough of County Cork, Munster, 15 miles North Northeast of Cork, on the Bride river. Area of parish, 13,995 acres. Population, 4,000 ; of town, 1,321. The town is neat and clean.

RATHCORMACK.—(or RATEGORMUCK.) Is a parish of County Waterford, Munster, 4 miles Southwest of Carrick-on-Suir. Area, 17,965 acres. Population, 2,498.

RATHDOWN CASTLE.—Is a ruin of the East coast of County Wicklow, Leinster, 2 miles South of Brayhead. RATHDOWN is the name given to two contigious baronies.

RATHDOWNEY.—Is a market town and parish of County Queens, Leinster, 6½ miles South Southeast of Borris-in-Ossory. Area of parish, 17,116 acres. Population, 6,756 ; of town, 1,414. It has Petty Sessions. There are seven Fairs held here Annually.

RATHDRUM.—Is a parish and market town of County Wicklow, Leinster, 8 miles West Southwest of Wicklow, on the Avonmore river. Area of parish, 5,798 acres. Population, 2,905 ; of town, 1,232.

RATHFARNHAM.—Is a large village and parish of County Dublin, Leinster, 3 miles South of Dublin. Area of parish, 2,782 acres. Population, 4,469. It contains numerous fine dwellings and demesnes, a fine Church and Rathfarnham Castle, the property of the Marquis of Ely, now converted into a Dairy.

RATHFRILAND.—Is a market town of County Down, Ulster, 9 miles East Northeast of Newry. Population, 2,183; chiefly employed in Linen weaving.

RATHGAR.—Is a village of County Dublin, Leinster, 2 miles South of Dublin. It has many Muslin, Calico and Print Works.

RATHGRAFF.—(or RATHGARVE.) Is a parish of County Westmeath, Leinster. Area, 6,024 acres. Population, 3,606. It comprises the town of Castle-Pollard.

RATHKEALE.—Is a market town and parish of County Limerick, Munster, 17 miles West Southwest of Limerick in the Deel river. Area of parish, 12,095 acres. Population, 8,293; which includes many descendants of German Protestants, (termed palatines) established here by the Southwell family whose seat, Castle Matress, is immediately South of the town. This town ranks second in the County, is very prosperous and has many fine buildings. Population, 4,201.

RATHKENNAN.—(or RATHKENNY.) There are several parishes by this name, as follows. viz : County Tipperary, Munster, 4 miles West Southwest of Holycross. Area, 787 acres. Population, 299. County Meath, Leinster, 5 miles Northwest of Slane. Area, 5,496 acres. Population, 2,179.

RATHLIN.—(RACHLIN or RAGHERY.) Is an Island off the coast of County Antrim, Ulster, forming a parish, situated in the North channel, 3 miles Northwest of Fairhead. Latitude, 55° 17′ 6″ North. Longitude, 6° 11′ West. Area, 3,398 acres. Population, 1,010. Like the Giant's Causeway on the opposite coast, it is of Basaltic formation. Among its numerous antiquities is a ruined Castle which in 1306 afforded a refuge to Robert Bruce.

RATHLIN-O'BIRNE.—Is a group of islets off Teelen Head, Ulster, on the North side of Donegal bay.

RATHMELTON.—Is a town of County Donegal, Ulster, on the West side of Lough Swilly, 6 miles West of Rathmullen. Population, 1,498.

RATHMINES.—Is a suburb of Dublin, on the South, 1½ miles South of Dublin Castle. Population, 2,429. It has a modern residence on the site of the battlefield where in 1649 the republicans totally defeated the forces of the Marquis of Ormonde.

RATHMOLYON —Is a parish of County Meath, Leinster, 2¾ miles West Northwest of Summerhill. Area, 9,783 acres. Population, 2,953; of whom 176 are in the village.

RATHMORE —There are several parishes by this name in Leinster, as follows, viz: County Carlow, 3½ miles North of Tullow. Area, 815 acres. Population, 323; County Kildare, 4½ miles East Northeast of Naas. Area, 7,756 acres. Population, 1,495; County Meath, 4 miles North Northeast of Athboy. Area, 5,345 acres. Population, 1,780. There is also a bog by this name in County Kerry, Munster, having an area of 1,371 acres.

RATHMULLEN.—Is a parish of County Down, Ulster, on Dundrum bay. Area, 3,369 acres. Population, 2,603. It comprises the town of Killough.

RATHNEW.—Is a maritime parish of County Wicklow, Leinster. Area, 8,641 acres. Population, 5,754; of village, 118. It comprises part of the town of Wicklow.

RATHOWEN.—Is a village of County Westmeath, Leinster, 12 miles North Northwest of Mullingar. Population, 550. It has a Church, School and Court House. There are two Fairs held here Annually.

RATHREAGH.—There are two parishes by this name. One is in County Longford, Leinster, 4 miles South of Edgeworthstown. Area, 4,023 acres. Population, 1,123; another is in County Mayo, Connaught, 3½ miles Northwest of Killala. Area, 4,164 acres. Population, 1,664.

RATHRONAU.—There are two parishes by this name in Munster. One is in County Limerick. Area, 18,117 acres. Population, 3,245. It comprises the village of Athea. The other is in County Tipperary, 3 miles North of Clonmel. Area, 2,641 acres. Population, 1,112.

RATHSALLAH.—Is a parish of County Wicklow, Leinster. Area, 1,776 acres. Population, 226.

RATHSARAN.—Is a parish of County Queens, 2 miles West of Rathdowney. Area, 2,291 acres. Population, 965.

RATHVILLY.—Is a village, parish and barony of County Carlow, Leinster, 10 miles East Northeast of Carlow. Area of parish, 9,212 acres. Population, 3,493; of village, 499.

RATOATH.—Is a village, parish and formerly parliamentary borough of County Meath, Leinster, 14 miles North Northwest of

Dublin. Area of parish, 9,331 acres. Population, 1,597; of village, 513.

RATTOO.—Is a parish of County Kerry, Munster, 6¼ miles West Southwest of Listowel. Area, 8,230 acres. Population, 3,860.

RAYMOCHY.—Is a parish of County Donegal, Ulster, on Lough Swilly. Area, 15,286 acres. Population, 5,733. It comprises the village of Manor-Conyngham.

RAYMUNTUDONY.—Is a parish of County Donegal, Ulster, 4½ miles Southwest of Dunfanaghy. Area, 12,163 acres. Population, 2,238.

REYMORE.—Is a parish of County Queens, Leinster, 6 miles West Northwest of Mountmellick. Area, 13,943 acres. Population, 2,916. Its Southern part comprises a part of the Slicbhloom mountains. Highest point, 1,676 feet.

REE.—(Lough.) Is a Lough formed by an expansion of the Shannon river, between the provinces of Leinster and Connaught. Length, 15 miles. Greatest breadth, 8 miles. Area, 42 square miles, or 26,880 acres. It contains many islets and receives the Inny river on the East.

REYNAGH.—Is a parish of County Kings, Leinster. Area, 8,827 acres. Population, 5,106. It comprises the town of Banagher. Here are the remains of several old baronial Castles and ecclesiastical edifices.

RICH HILL.—Is a market town of County Armagh, Ulster, 4 miles East Northeast of Armagh. Population, 752. It has quite a Linen trade.

RINGROVE.—Is a maritime parish of County Cork, Munster, 2 miles South of Kinsale. Area, 9,240 acres. Population, 5,455. Here are the ruins of Ringrove Castle, which gives the title of Baron to Lord Kinsale.

RINGSEND.—Is a suburb of Dublin, ½ mile East of Dublin Castle. It adjoins Irishtown and forms one of the lowest quarters of Dublin.

ROARING WATER BAY.—Is a bay of County Cork, Munster, extending inland for 9 miles behind Cape Clear.

ROBE.—Is a river of County Mayo, Connaught, rising near Clare and after a westward course of 26 miles flows into Lough Mask, 2 miles West of Ballinrobe.

ROBEEN.—Is a parish of County Mayo, Connaught, 2 miles

Northwest of Hollymont. Area, 10,907 acres. Population, 3,544.

ROBERTSTOWN.—There are two parishes by this name. One is in County Meath, Leinster, 4 miles West of Drumconrath. The other or (CASTLE ROBERT,) is in County Limerick, Munster, 4½ miles West of Askeaton. Area, 5,906 acres. Population, 2,314.

ROCHESTOWN.—There are two parishes by this name. One is in County Tipperary, Munster, 3 miles East Southeast of Cahir. Area, 1,063 acres. Population, 488. The other, (or BALLY WILLIAM) is in County Limerick, Munster, 2 miles North Northwest of Six-mile-Bridge. Area, 1,165 acres. Population, 273. It is also the name of a hamlet in County Dublin, Leinster, 1½ miles East Southeast of Dundrum.

ROSBERCON.—(or ROSEBERCON.) Is a village and parish of County Kilkenny, Leinster, on the West side of the Barrow river, opposite New Ross, of which it is a suburb. Area, 1,705 acres. Population, 1,538. It has large Stores, Quays, Colcomb Distillery and picturesque remains of a Monastery.

ROSCOMMON.—Is an inland County of Connaught; having the Counties, Longford and Westmeath on the East, from which it is separated by the Shannon river; the Counties, Galway and Mayo on the West, separated by the Suck river; the Counties, Sligo and Leitrim on the North. Area, 607,691 acres; of which 130,300 are waste. Population in 1851, 174,432. The surface is mostly undulating, being mountainous in the North and flat in the East. Soil generally fertile and pastures fine. The chief productions are Oats, Potatoes and Wheat. Land near the towns rents at from £3 to £4 per acre, but average rent elsewhere is less than 20 Shillings. The manufactures have much declined. The County is divided into 9 baronies. It sends two members to the House of Commons. The principal towns are Roscommon, Castlereagh, Boyle, Strokestown and a part of Athlone.

ROSCOMMON.—Is a market town and parish, and formerly a parliamentary borough and capital of County Roscommon, Connaught, 17 miles North Northwest of Athlone. Area of parish, 9,819 acres. Population, 8,191; of town, 3,439. The town is poorly built, the principal buildings being the parish Church, Roman Catholic Chapel, Jail, new Court House, some Manufactories, etc. Markets, Saturday. It gives title of Earl to the Dillon family.

ROSCREA.—Is a market town and parish of County Tipperary,

7 miles West of Borris-in-Ossory. Area of parish, 4,829 acres. Population, 5,275. It is well situated but poorly built, having some good Public Buildings, Churches, Manfactories, etc. Markets, Monday and Thursday.

ROSENALLIS.—Is a parish of County Queens, Leinster. Area, 41,119 acres. Population, 8,505 ; of village 239, with part of the town of Mountmellick.

ROSS.—There are numerous localities by this name, as follows, viz: an island in the lower lake of Killarney, County Kerry, Munster, 2 miles South of Killarney. Area, 100 acres; on it are the remains of a strong Castle ; a bog of County Queens, Leinster, 3 miles West Northwest of Maryboro'. Area, 3,007 acres ; a parish of County Galway, Connaught, 12½ miles Northwest of Oughterard. Area, 59,651 acres. Population, 4,804. It comprisse part of Loughs, Mask and Corrib. The Devils mountain in this parish rises to the height of 2,131 feet ; a harbor of County Mayo, Connaught, on the East side of Broadhaven ; a bar of County Galway; and a village in County Clare. It is also the name of a small Lough.

ROSS.—(or ROSSCARBERY.) Is a market town, parish and Episcopal See of County Cork, Munster. The town is situated 7 miles West Southwest of Clonakilty. Area, 13,350 acres. Population, 8,839 ; of town, 1,530. It is poorly built, and has several good buildings, Churches, etc. The See comprises 32 parishes in the same County, and is united to the diocese of Cork and Cloyne.

ROSS.—(NEW.) Is a parliamentary and municipal borough, river port, town and parish of County Wexford, Leinster, on the Barrow river, 13 miles North Northeast of Waterford and 17 miles West Southwest of Enniscorthy. Area of parish, 4,922 acres. Population in 1851, 9,131. Area of parliamentary borough, 390 acres. Population in 1851, 7,778. It is well built and enclosed by old walls, and has a Quay 650 feet in length. It contains many fine buildings and has an extensive trade. Exports, Corn, Flour, Wool, Butter, Cattle and Bacon. It sends one member to the House of Commons. It gives title of Earl to the Parsons family.

ROSSDROIT.—Is a parish of County Wexford, Leinster, 4 miles West Southwest of Enniscorthy. Area, 8,866 acres. Population, 2,258.

ROSSDUFF.—Is a parish of County Waterford, Munster, 2¾ miles Northwest of Dunmore. Area, 197 acres. Population, 111.

ROSSINVER.—Is a parish of Counties, Sligo and Leitrim, Connaught. Area, 48,843 acres. Population, 28,130. It comprises the village of Kinlough.

ROSSMERE.—(or ROSSMIRE.) Is a parish of County Waterford, Munster. Area, 8,161 acres. Population, 2,866. It comprises the town of Kilmacthomas.

ROSSORY.—Is a parish of County Fermanagh, Ulster. Area, 7,654 acres. Population, 3,846. It comprises part of the town of Enniskillen.

ROSTREVOR.—Is a small maritime town of County Down, Ulster, on the North side of Carlingford bay, 8 miles East Southeast of Newry. Population, 683. It is finely situated and much resorted to by bathers. There are seven Fairs held here Annually.

RUSH.—Is a seaport and market town of County Dublin, Leinster, situated on a headland projecting into the Irish sea, 14 miles Northeast of Dublin. Population, 1,603. It has a small harbor and coast-guard station.

RUTLAND ISLAND.—Is an island off the coast of County Donegal, Ulster, immediately East of North Arran.

SAINTFIELD.—(or TULLAGHNANOEVE.) Is a market town and parish of County Down, Ulster, 9 miles South Southeast of Belfast. Area of parish, 13,334 acres. Population, 6,247 ; of town, 909. It has a thriving trade in Linens, Calicoes, Corduroys and other fabrics.

SALLINS.—Is a village of County Kildare, on the Grand Canal, 7½ miles Northeast of Newbridge. Population, 392.

SALTEE ISLANDS.—Are two small islands and a group of rocks off the coast of County Wexford, Leinster. The large island lies about 15 miles East of Hook-head, and is 1 mile in length. A vessel showing a fixed double light is stationed 3¼ miles West Southwest of this island.

SANDY MOUNT.—Is a village of County Dublin, Leinster, 2¼ miles South Southeast of Dublin, on Dublin bay. It is much frequented for sea-bathing.

SCARIFF.—Is a market town of County Clare, Munster, on the Scariff river, 8 miles North Northwest of Killaloe. Population, 656.

SCRABBY.—(or BALLYMACALLENY.) Is a parish of County Cavan, Ulster, 3½ miles South Southeast of Armagh. Area of parish, 6,661 acres. Population, 2,836 ; of village, 170.

SEAGOE.—(or SEGOE.) Is a parish of County Armagh, Ulster. Area, 10,982 acres. Population, 11,094. It comprises a part of the town of Portadown.

SEAPATRICK.—Is a parish of County Down, Ulster. Area, 7,583 acres. Population, 9,528. It comprises the town of Banbridge.

SESKINAN.—Is a parish of County Waterford, Munster, 6 miles Northeast of Cappoquin. Area, 16,877 acres. Population, 3,210.

SHANAGOLDEN.—Is a parish of County Limerick, Munster, 5 miles Southwest of Askeaton. Area, 4,233 acres. Population, 2,716 ; of village, 548.

SHANDRUM.—Is a parish of County Cork, Munster, 2 miles West Northwest of Charleville. Area, 13,451 acres. Population, 5,164.

SHANKILL.—There are several parishes by this name, as follows, viz : County Armagh, Ulster. Area, 6,514 acres. Population, 9,350. It comprises the town of Lurgan ; County Roscommon, Connaught, 1¾ miles West of Elphin. Area, 6,611 acres. Population, 2,626 ; (or ST. KILL,) County Kilkenny, Leinster, 2 miles North of Gowran. Area, 6,489 acres. Population, 2,506. It is also the name of a hamlet in County Dublin.

SHANNON.—Is the principal river of Ireland. It rises in a pond called the Shannon Pot, near the base of Cuilcagh mountain, County Cavan, and flows generally South and empties into the Atlantic Ocean, after a course of 224 miles. It divides Connaught from Leinster, and traverses the Northern part of Munster, separating County Clare from Counties, Tipperary, Limerick and Kerry. It also, by its widening, forms Loughs, Allen, Baffin, Ree and Derg. It is navigable to near its source, although it is obstructed in many places by rapids and shallows, to overcome which, large sums have been expended. Its affluents are the Boyle, Suck and Fergus rivers on the West, and the Inny. Brosna, Mulkerna and Maig rivers on the East. It is tidal for the East third of its course, and may be ascended by vessels of 400 tons to Limerick. It is connected with Dublin by the Grand Canal from Shannon harbor near Banagher, and by the Royal Canal, which joins it at Farmonbarry, near Longford. It gives title of Earl to the Boyle family.

SHANNON BRIDGE.—Is a village and fortified post on the

Shannon river, County Kings, Leinster, 2½ miles North Northeast of Banagher. Population, 398.

SHANNON HARBOR and SHANNON GROVE.—Are villages lower down the river.

SHANRAHAN.—Is a parish of County Tipperary, Munster. Area, 24,923 acres. Population, 7,398. It comprises the town of Clogheen.

SHEELIN.—(Lough.) Is a lake partly in Counties, Meath and Westmeath, Leinster, but chiefly in County Cavan, Ulster, 5 miles East of Granard, 5 miles in length.

SHERCOCK.—(or Killan.) Is a parish of County Cavan, Ulster, 9½ miles Southeast of Cotehill. Area, 8,221 acres. Population, 5,544; of village 511.

SHILLELAGH.—Is a village of County Wicklow, Leinster, on the Shillelagh, an affluent of Slaney, 9 miles East Southeast of Tullow. Population, 186. It has a handsome Church and other Buildings.

SHINRONE.—Is a market town and parish of County Kings, 6 miles West Northwest of Roscrea. Area of parish, 4,869 acres. Population, 2,563; of town, 1,054. There are remains of several Castles here.

SIX-MILE-BRIDGE.—Is a market town of County Clare, Munster, on the Augarnee river, 9 miles Northwest of Limerick. Population, 848. It has a Court House, Bridewell and Market House. Is also the name of a market town of County Limerick, 10 miles South Southeast of Limerick. Population, 174.

SIX-MILE-CROSS.—Is a village of County Tyrone, Ulster, on the Clogphin river, 8 miles West Southwest of Pomeroy. Population, 600. There are twelve Fairs held here Annually.

SKERRIES.—Is a fishing town of County Dublin, Leinster, 17 miles North Northeast of Dublin, on a headland ¾ miles East of the Dublin and Drogheda Railroad. Population, 2,417. It is clean and well built and has a good pier. It is also the name of an islet in the Irish sea off the Northwest coast of Anglesey. It has a lighthouse. Latitude, 53° 25′ 3″ North. Longitude, 4° 36′ 5″ West. Elevation, 117 feet.

SKERRY.—Is a parish of County Antrim, Ulster, 3 miles East Northeast of Broughshane. Area, 26,176 acres. Population, 5,349.

SKIBBEREEN.—Is a market town of County Cork, Munster, 40 miles Southwest of Cork, on the river Ilen. Population, 4,715.

It is brisk and thriving, and has a good Church, several Schools, Bridewell, etc. In the vicinity are the remains of several feudal Castles and a Monastery.

SKIRTS.—(or DERG.) Is a parish of County Tyrone, Ulster. Area, 14,286 acres. Population, 5,799. It comprises the town of Castle-Derg.

SKREEN.—(or SKRYNE.) Is a barony, parish and village of County Meath, Leinster. The village is situated 6 miles Southeast of Navan. Population, 225. There are remains of Ecclesiastical edifices. Area of parish, 4,251 acres. Population, 1,156. It is also the name of a parish of County Sligo, Connaught, 10½ miles West Southwest of Sligo. Area, 13,237 acres. Population, 4,103. It is also a parish of County Wexford, Leinster, 6 miles Northeast of Wexford. Population, 820.

SKULL.—Is a parish of County Cork, Munster, 11 miles West of Skibbereen. Area, 37,923 acres. Population, 17,314; of village, 452.

SLANE.—Is a market town and parish of County Meath, Leinster, on the Boyne river, 8 miles West of Drogheda. Area, 5,947 acres. Population, 2,510; of town, 555.

SLANES.—Is a parish of County Down, Ulster, 3 miles Northeast of Portaferry. Area, 946 acres. Population, 556.

SLANEY.—Is a river of Leinster, rising in County Wicklow, flows South through the Counties, Carlow and Wexford, and empties in Wexford harbor after a course of 60 miles. Tidal for 10 miles and navigable to Enniscorthy.

SLIEVE-BEG.—Is a mountain of County Down, Ulster, 2½ miles West Southwest of Newcastle. Elevation, 2,384 feet.

SLIEVE-CAR.—Is a mountain of County Mayo, Connaught, between Lough Coon and Blacksod bay. Elevation, 2,368 feet.

SLIEVE-DONARD.—Is a mountain of County Down, Ulster, on the Southwest side of Dundrum bay, 2 miles Southwest of Newcastle. Elevation, 2,796 feet.

SLIGO.—Is a maritime County of Connaught; having the County Leitrim on the East, the Counties, Roscommon and Mayo on the South, and the bays of Killala, Sligo and Donegal on the Northwest and North. Area, 461,753 acres; of which 115,438 acres are waste. Population in 1851, 128,511. Surface greatly diversified; soil fertile in some parts, and watered by the Arrow and Moy rivers, the latter forming the West boundary, and by

Loughs, Arrow, Gill and a part of Garra. Agriculture improved, the principal crops being Oats, Potatoes and Wheat. Estates large, but farms are small. Coarse Linens and Woolens are manufactured here. The fisheries employ about 2,000 persons. The County is subdivided into 6 baronies and 41 parishes in the diocese of Achonry and Elphin. This County sends two members to the House of Commons.

SLIGO.—Is a parliamentary and municipal borough, and seaport town and capital of County Sligo, Connaught, at the head of Sligo bay, and at the mouth of the Sligo river, 69 miles South Southwest of Londonderry. Area of parliamentary borough, 3,001 acres. Population, 11,511; of town, 2,046. The town has many handsome edifices, two Churches, a large Roman Catholic Chapel, Monastery, County Court House, Prison, etc. The harbor is much improved, and vessels of 300 tons can approach the town. It has several Flour Mills and Breweries. The borough sends one member to the House of Commons. It gives the title of Marquis to the Brown family.

SLIGO BAY.—Is an inlet of the Atlantic, South of Donegal bay, extending inland for 12 miles, and having a breadth at its entrance of 6 miles.

SOLLOGHODMORE.—Is a parish of County Tipperary, Munster, 3½ miles Northwest of Tipperary. Area, 6,657 acres. Population, 2,852.

SORRELL HEAD.—Is a mountain of County Wicklow, Leinster, 4 miles Southeast of Blessington. Elevation, 1,915 feet.

STEWARTSTOWN.—Is a market town of County Tyrone, Ulster, 7 miles North Northeast of Dungannon. Population, 1,082.

STILLORGAN.—Is a village of County Dublin, Leinster, 5 miles Southeast of Dublin. Population, 611. It gives the title of Baron to the Allen family.

STRABANE.—Is a municipal borough and market town of County Tyrone, Ulster, on the Mourne river, and near its confluence with the Finn and Foyle rivers. Population, 3,611. Its trade is facilitated by a canal from it to where the Foyle river becomes navigable for barges of 40 tons. Adjoining the town is a Salmon Fishery.

STRADBALLY.—Is a market town and parish of County Queens, Leinster, on the Strad river, 7 miles East Southeast of Maryboro'. Area of parish, 2,467 acres. Population, 2,588; of town, 1,682. Near it is Rockley Park, formerly the residence of the Earl of Roden.

STRADBALLY.—Is a parish of County Waterford, Munster, 8 miles East Northeast of Dungarven. Area, 10,917 acres. Population, 4,419; of village, 814.

STRADBALLY.—Is a parish of County Galway, Connaught, 3¼ miles South Southeast of Oranmore. Area, 4,618 acres. Population, 1,264; of village, 280. It is also the name of a parish of County Kerry, Munster, 15 miles West of Tralee. Area, 4,103 acres. Population, 1,202; of village, 336. It is situated on the shore of Bandon bay.

STRADONE.—Is a village of County Cavan, Ulster, 5 miles East Southeast of Cavan. Population, 322. There are eight Fairs held here Annually.

STRAFFAN.—Is a parish of County Kildare, Leinster, 17 miles West Southwest of Dublin. Area, 2,286 acres. Population, 834.

STRAID.—(or TEMPLEMORE.) Is a parish of County Mayo, Connaught, 4 miles South of Foxford. Area, 9,465 acres. Population, 4,251. Here are the remains of an Abbey and an Old Castle.

STRANGFORD.—Is a seaport and market town of County Down, Ulster, 6 miles East Northeast of Downpatrick. Population, 571. It employs many vessels in the fisheries. Adjoining are Castleward and four Forts. It gives the title of Viscount to the Smyth family.

STRANGFORD.—(LOUGH.) Is a large lake or inlet of the sea, between Lough Belfast and Dundrum bay, 15 miles in length. It is shallow.

STRANOLAR.—Is a market town and parish of County Donegal, Ulster, on the Finn river, ½ mile Northeast of Ballybofey. Area of parish, 15,508. Population, 4,944; of town, 385. It comprises the town of Ballybofey.

STRATFORD UPON SLANEY.—Is a small manufacturing town of County Wicklow, Leinster, on the Slaney river, 13 miles South Southwest of Blessington. Population, 618.

STROKESTOWN. Is a market town of County Roscommon, Connaught, 6½ miles South Southeast of Elphin. Population, 1,611 The demesne of Lord Hartland (Bawn) is adjacent.

SUIR Is a river of Munster, rising in the Devil's Bit mountains flowing South for 100 miles, emptying into Waterford harbor.

SWADLINBAR.—Is a small town of County Cavan, Ulster, 8 miles Northwest of Ballyconnell. Population, 492.

SWILLY.—(Lough.) Is an inlet of the Atlantic on the North coast of County Donegal, Ulster, extending inland for 25 miles, where it receives the Swilly river. There is a Light House on Tannet Point, on the West side of its entrance. Latitude, 55° 16' 33" North. Longitude, 7° 38' West.

SWINEFORD.—Is a market town of County Mayo, Connaught, 15½ miles East Northeast of Castlebar. Population, 1,016.

SWORDS.—Is a market town and parish of County Dublin, Leinster, 8 miles North Northeast of Dublin, on Swords river. Area, 9,675 acres. Population, 3,638; of town, 1,788. It has many fine buildings, but the town is much decayed.

TAGHBOY.—(or Taughboy.) Is a parish of County Roscommon, Connaught, 5¾ miles Southeast of Athleague. Area, 13,997 acres. Population, 3,825.

TAGHEEN.—(or Taugheen.) Is a parish of County Mayo, Connaught, 2¾ miles North Northeast of Hollymount. Area, 6,837 acres. Population, 3,084.

TAGHMACONNELL.—Is a parish of County Roscommon, Connaught, 4½ miles North Northeast of Ballinasloe. Area, 18,876 acres. Population, 4,807.

TAGHMON.—Is a disfranchised parliamentary borough and parish of County Wexford, Leinster, 7½ miles West Southwest of Wexford. Area, 10,125 acres. Population, 3,737; of town, 1,303. It is poorly built. It is also the name of a parish of County Westmeath, Leinster, 6 miles North Northeast of Mullingar. Area, 3,453 acres. Population, 958.

TALLAGHT.—Is a parish of County Dublin, Leinster, 5 miles Southwest of Dublin. Area, 21,868 acres. Population, 4,921; of village, 348.

TALLOW.—Is a market town and parish of County Waterford, Munster, 42 miles North Northwest of Youghal. Area of parish, 5,027 acres. Population, 4,867; of town, 2,969. It has a handsome Church and large Roman Catholic Chapel.

TALLOW-BRIDGE.—Is a village ½ mile North Northeast of Tallow. Population, 258.

TAMLAGHT.—There are several parishes by this name, as follows, viz: County Tyrone, Ulster, 3¼ miles South Southeast of Moneymore. Area, 4,955 acres. Population, 3,006; (who are engaged in the manufacture of Linen.) There is a large Druidical Altar here; County Londonderry, Ulster, 6 miles Northeast of Maghera. Area, 17,402 acres. Population, 6,616 ; (or O'CRILLY,) County Londonderry, Ulster. Area, 16,840 acres. Population, 16,849. It comprises part of the town of Portglenone.

TANEY.—(TAWNEY or CHURCH-TOWN.) Is a parish of County Dublin, Leinster, 2½ miles South Southeast of Dublin. Area, 4,563 acres. Population, 3,848. Is also the name of a village of County Donegal, Ulster, 6¾ miles West of Millford. Population, 128.

TANKARDSTOWN.—There are two parishes by this name. One is in Counties, Kildare and Queens, Leinster, 3½ miles South Southeast of Athy. Area, 8,350 acres. Population, 1,914 ; another is in County Limerick, Munster, 1 mile West of Kilmallock. Area, 1,710 acres. Population, 660.

TARA.—Is a parish of County Meath, Leinster, 2 miles West of Skreen. Area, 3,364 acres. Population, 586. The hill of Tara was in remote antiquity the chief seat of the Irish Monarchs, and from it was originally brought the famous stone long used in the Coronation of the Scottish Kings at Scone, and now in the chair of Edward the Confessor at Westminster.

TARBERT.—Is a seaport town of County Kerry, Munster, on the Shannon river, 4 miles West Northwest of Glin. Population, 1,024.

TARMONBARRY.—(or TERMONBARRY.) Is a parish of County Roscommon, Connaught, 8 miles East Southeast of Strokestown. Area, 9,295 acres. Population, 4,279.

TARTARAGHAN.—Is a parish of County Armagh, Ulster, 4 miles North Northeast of Longhgall. Area, 11,612 acres. Population, 7,313.

TASHINNY.—(or TAGHSHINNY.) Is a parish of County Longford, Leinster, 3 miles Northeast of Ballymahon. Area, 4,881 acres. Population, 2,333.

TEDAONET.—(or TEDONAGH.) Is a parish of County Monaghan, Ulster, 4 miles Northwest of Monaghan. Area, 26,502 acres. Population, 11,645.

TEHALLON.—(or TYHALLON.) Is a parish of County Monaghan, Ulster, 3½ miles East Northeast of Monaghan. Area, 5,949 acres. Population, 3,806 ; partly employed in Linen Weaving.

TEIGHSHINOD.—(or TAGHSHEENOD.) Is a parish of County

Longford, Leinster, 3¼ miles Northeast of Ballymahon. Area, 5,713 acres. Population, 2,533.

TEMPLEBOY,—Is a maritime parish of County Sligo, Connaught, 2 miles Southeast of Dunmore. Area, 9,113 acres. Population, 3,812.

TEMPLEBREADY.—Is a parish of County Cork, Munster, 6 miles South Southeast of Passage. Area, 2,654 acres. Population, 1,613.

TEMPLEBREDIN.—Is a parish of Counties, Tipperary and Limerick, Munster, 2½ miles North of Emly. Area, 2,455 acres. Population, 1,457.

TEMPLECARNE.—Is a parish of Counties, Fermanagh and Donegal, Ulster. Area, 45,870 acres. Population, 5,934. It comprises part of the town of Pettigoe.

TEMPLECORRAN.—Is a parish of County Antrim, Ulster, 4½ miles Northeast of Carrickfergus. Area, 4,744 acres. Population, 1,428. The Church of Templecorran, now ruined, was the first benefice to which Dean Swift was appointed.

TEMPLECRONE.—Is a maritime parish of County Donegal, Ulster. Area, 52,921 acres. Population, 9,842. It comprises the town of Dunglo.

TEMPLEDERRY.—Is a parish of County Tipperary, Munster, 7½ miles Southeast of Nenagh. Area, 6,990 acres. Population, 2,032.

TEMPLEKELLY.—(or TEMPLEJEHALLY.) Is a parish of County Tipperary, Munster. Area, 10,039 acres. Population, 4,259. It comprises the town of Ballina.

TEMPLEMARTIN.—Is a parish of County Cork, Munster, 5½ miles North of Bandon. Area, 7,515 acres. Population, 2,362.

TEMPLEMARTIN.—(or St. Martin.) Is a parish of County Kilkenny, Leinster, 2 miles East of Kilkenny. Area, 782 acres. Population, 306.

TEMPLEMICHAEL.—There are several parishes by this name, as follows, viz : County Longford, Leinster. Area, 9,115 acres. Population, 8,484. It comprises the town of Longford ; County Waterford, Munster, 2¾ miles North Northwest of Youghal. Area, 8,216 acres. Population, 2,994 ; County Tipperary, Munster, 11 miles East Northeast of Clonmel ; (or DEDUAGH,) County Cork, Munster, 2½ miles East Northeast of Innishannon. Area, 2,064 acres. Population, 711.

TEMPLEMORE.—Is a town and parish of County Tipperary, Munster, near the foot of the Devil's Bit mountains, 8 miles North of Thurles. Area of parish, 8,472 acres. Population, 5,966; of town 3,685. It is well built and has many good buildings. It is also the name of a parish of County Londonderry, Ulster. Area, 12,616 acres. Population, 20,379. It comprises the greater part of the city of Londonderry.

TEMPLENEIRY.—Is a parish of County Tipperary, Munster, 3½ miles from Tipperary. Area, 12,840 acres. Population, 3,700. The Galtee mountains are in the parish. Highest elevation, 2,588 feet.

TEMPLENOE.—Is a parish of County Kerry, Munster, 5½ miles West Southwest of Kenmare. Area, 32,428 acres. Population, 4,189.

TEMPLENOE.—(or Lisnavilla.) Is a parish of County Tipperary, Munster, 3 miles Northeast of Tipperary. Area, 2,730 acres. Population, 1,154.

TEMPLEOUTRAGH.—(or Upperchurch.) Is a parish of County Tipperary, Munster, 8 miles Northwest of Holycross. Area, 12,903 acres. Population, 3,144.

TEMPLEPATRICK—Is a parish of County Antrim, Ulster, 5½ miles East Southeast of Antrim. Area, 14,191 acres. Population, 3,559; of village, 194.

TEMPLEPORT.—Is a parish of County Cavan, Ulster, 4 miles West Southwest of Ballyconnel. Area, 42,172 acres. Population, 12,100.

TEMPLEROBIN.—Is a parish of County Cork; consisting of Spike and Hawlbowline islands, in Cork harbor, and comprises part of Great Isle, with the chief part of Queenstown. Area, 3,594 acres. Population, 7,391.

TEMPLESHAMBO.—(or Templeshanbough.) Is a parish of County Wexford, Leinster, 6 miles West Southwest of Newtownbarry. Area, 19,516 acres. Population, 6,907.

TEMPLESHANNON.—Is a parish of County Wexford, Leinster. Area, 4,903 acres. Population, 3,232. It comprises part of the town of Enniscorthy.

TEMPLETENNY.—Is a parish of County Tipperary, Munster, 5 miles West of Clogheen. Area, 18,182 acres. Population, 6,907.

TEMPLETOGHER.—Is a parish of County Galway, Connaught, 3½ miles Northwest of Ballymoe. Area, 13,706 acres. Population, 4,976.

TEMPLETOWN.—Is a parish of County Wexford, Leinster. 2¼ miles West Southwest of Fethard. Area, 4,157 acres. Population, 1,426. It is also the name of a village of County Louth, Leinster, 3½ miles South Southeast of Carlingford.

TEMPLETRIME.—Is a parish of County Cork, Munster, 5 miles Southwest of Kinsale. Area, 4,784 acres. Population, 2,149.

TEMPLETUOHY.—(or BALLINSIN.) Is a parish of County Tipperary, Munster, 5 miles East of Templemore. Area, 8,461 acres. Population, 3,194 ; of village, 393, near which are the ruins of Lisdallen Castle.

TEMPLEUDIGAN.—(or ST. PETERS.) Is a parish of County Wexford, Leinster, 5 miles North Northwest of New Ross. Area, 8,177 acres. Population, 2,151.

TEMPO.—Is a neat village of County Fermanagh, Ulster, on the Tempo river, 8 miles East Northeast of Enniskillen. Population, 422.

TERMONAMUNGAN.—Is a parish of County Tyrone, Ulster, 3¾ miles Southwest of Castle-Derg. Area, 45,399 acres. Population, 7,561.

TERMONEENY.—Is a parish of County Londonderry, Ulster, 3½ miles Southeast of Maghera. Area, 4,774 acres. Population, 2,539.

TERMONFECKAN.—(or TORFECKAN.) Is a parish and village of County Louth, Leinster, on the East coast, 2 miles South Southwest of Clogher. Area of parish, 6,382 acres. Population, 3,344 ; of village, 412. It is much resorted to for sea bathing.

TERMONMAQUIRK.—Is a parish of County Tyrone, Ulster, 4 miles West of Pomeroy. Area, 41,079 acres. Population, 12,098.

TERRYGLASS.—Is a parish of County Tipperary, Munster, 5 miles Northwest of Borris O'Kane. Area, 9,762 acres. Population, 2,953.

TESSARAGH.—(or TAUGHSRARA.) Is a parish of County Roscommon, Connaught, 2 miles South Southwest of Athleague. Area, 8,482 acres. Population, 3,356.

TESSAURAN.—(or KILGALLY.) Is a parish of County Kings, Leinster, 1½ miles Northwest of Cloghan. Area, 7,317 acres. Population, 2,029.

THURLES.—Is a market town and parish of County Tipperary, Munster, 21 miles Northeast of Tipperary, on the Suir river. Area, 8,269 acres. Population of town in 1851, 5,921 ; of parish, 10,284. It has a Roman Catholic Chapel, Palace and Episcopal Palace, College, etc.

TICKMACREVAN.—Is a parish of County Antrim. Area, 20,507 acres. Population, 4,444. It comprises the town of Glenarm.

TIMOLEAGUE.—Is a decayed market town and parish of County Cork, Munster, on Courtmacksherry Bay, 7 miles South Southwest of Bandon. Area of parish, 2,873 acres. Population, 1,686 ; of town, 635.

TINEHELY.—Is a market town of County Wicklow, Leinster, 6½ miles Southwest of Aughrim. Population, 640. It has been rebuilt since its destruction in 1798, by Earl Fitzwilliam, whose seat Coolattin Park, is in the vicinity.

TINTERN.—(or KINNEAGH.) Is a parish of County Wexford, 3 miles North of Fethard. Area, 6,863 acres. Population, 2,680. Here are remains of an Abbey.

TIPPERARY.—Is an inland County of Munster. Area, 1,061,730 acres ; of which 191,706 are uncultivated. Population in 1851, 331,487. Surface diversified and in some parts mountainous soil generally is of extraordinary fertility, yielding large crops, Agriculture is much improved but farms are small. The Knockmeledown and Devils Bit are the principal mountains ; the Suir is the principal river. This County is divided into North and South Ridings, 12 baronies and 193 parishes in the dioceses of Cashel, Emly, Killaloe and Lismore. It sends 2 members to the House of Commons.

TIPPERARY.—Is the capital of County Tipperary, on the Arra river, an affluent of the Suir, 25 miles Southeast of Limerick. Population in 1851, 6,130. It is situated in a fertile track and is well built and thriving. It contains many handsome Churches, Public and Charitable Institutions, etc.

TOBBER.—Is a parish of County Dublin, Leinster, 1¼ miles Northeast of Dunlavin. Area, 1,434 acres. Population, 576. There is a village of same name in County Tipperary, Munster, 2½ miles Southwest of Clonmel, near the Suir river. Population, 149. Is also name of a hamlet of County Kings, Leinster, 3 miles Northwest of Clare.

TOBBERAHEENA.—Is a village of County Tipperary, Munster, on the Suir river, 2½ miles Southwest of Clonmel. Population, 453.

TOBBERCURRY.—Is a market town of County Sligo, 19 miles East Southeast of Ballina. Population, 783.

TOBBERMORE.—Is a small town of County Londonderry, Ulster, 5 miles Northwest of Magherafelt. Population, 525.

TOMFINLOUGH.—Is a parish of County Clare, Munster. Area, 6,736 acres. Population, 4,401. It comprises the town of Newmarket-on-Fergus.

TOMGRANEY.—(or TOMGRINI.) Is a parish of County Clare, Munster, 7 miles North Northwest of Killaloe. Area, 14,181 acres. Population, 6,113; of village, 371.

TOMREGAN.—Is a parish chiefly of County Cavan, Ulster. Area, 10,677 acres. Population, 4,212. It comprises part of the town of Ballyconnel.

TOOM.—(or TOOMVERIG.) Is a parish of County Tipperary, Munster, 6 miles Northwest of Tipperary. Area, 12,278 acres. Population, 4,277.

TOOMAVARRA.—Is a village of County Tipperary, Munster, 7 miles East Southeast of Nenagh. Population, 885.

TOOMB.—(or TOOME.) Is a parish of County Wexford, Leinster, 5½ miles Southwest of Gorey. Area, 6,979 acres. Population, 4,087.

TORY ISLAND.—Is an island off the Northwest coast of County Donegal, Ulster, 5 miles North Northwest of Innisboffin, with a Light House. Latitude, 55° 5' North. Longitude, 8° 15' West. Area, 785 acres. Population, 700.

TOWMORE.—(or TUOMORE.) Is a parish of County Mayo, Connaught. Area, 6,787 acres. Population, 3,744. It comprises the town of Foxford.

TRACTON.—Is a parish of County Cork, Munster, 3 miles South of Carrigaline. Area, 5,862 acres. Population, 2,959; of village, 115.

TRALEE.—Is a parliamentary and municipal borough, seaport town and parish of County Kerry, Munster, on the Lee river, 59 miles North Northwest of Cork. Area of parish, 4,605 acres. Population, 12,564. Area of parliamentary borough, 546 acres. Population in 1851, 9,916. It is well built, lighted and cleaned, and is rapidly increasing. It has many public edifices. Vessels of 300 tons can approach the town by means of a Ship Canal. Markets, Tuesdays and Saturdays. There are five Fairs held here Annually. It is the seat of County Assizes and Quarter Sessions, and head of a Poor Law Union. The borough sends one member to the House of Commons.

TRALEE BAY.—Is the Southern estuary of the Shannon river, and is 15 miles in length. Ballyheigue town and bay are on its North shore.

TRIM.—Is a disfranchised parliamentary borough, market town, parish and capital of County Meath, Leinster, on the Boyne river, 25 miles Northwest of Dublin. Area of parish, 13,426 acres. Population, 6,314; of town, 2,269. It is very old, and was formerly enclosed by walls. It has many public edifices, Schools, etc. Trim Castle was founded in the reign of Henry II. There is a handsome pillar erected in honor of the Duke of Wellington, whose birthplace, the demesne of Dangan, is situated South of the town.

TRORY.—(or ST. MICHAEL'S TRORY.) Is a parish of County Fermanagh, Ulster, 3½ miles North of Enniskillen. Area, 8,069 acres. Population, 2,028.

TUAM.—Is an Episcopal town and parish of County Galway, Connaught, 19 miles North Northeast of Galway, on the Clare river. Area of parish, 25,026 acres. Population, 13,425; of town, or city, 6,034. The town consists of five principal streets, Market Place, and some squalid straggling thoroughfares. It has a Roman Catholic and Protestant Cathedrals, and two Episcopal Palaces, the Roman Catholic College of St. Jarlath, Diocesan and other Public Schools, Court House, Bridewell, etc. It has Linen and Canvas Manufactories. It has a large retail trade. It was until 1839 the See of a Protestant Archbishop, but now nineteen-twentieths of the population are Roman Catholics.

TUBBRID.—Is a parish of County Tipperary, Munster, 4 miles South Southwest of Cahir. Area, 12,573 acres. Population, 4,874. There is also a parish of this name in County Kilkenny, Leinster, 3½ miles East Southeast of Pilltown. Population, 241.

TULLA.—(or TULLOH.) Is a market town and parish of County Clare, Munster, 9½ miles Northeast of Ennis. Area of parish, 24,532 acres. Population, 8,748; of town, 1,217.

TULLAGH.—Is a parish of County Cork, Munster. Area, 5,349 acres. Population, 3,690. It comprises the part of Baltimore, the island of Innisherkin.

TULLAGHANBROGUE.—Is a parish of County Kilkenny, Leinster, 4 miles Southwest of Kilkenny. Area, 3,187 acres. Population, 1,078.

TULLAGHANOGE.—Is a parish of County Meath, Leinster, 2½ miles Southeast of Athboy. Area, 1,415 acres. Population, 178.

TULLAGHLEASE.—(or TULLILEASE.) Is a parish of County Cork, Munster, 7 miles North Northeast of Newmarket. Area, 8,292 acres. Population, 3,278.

TULLAGHNISKEN.—Is a parish of County Tyrone, Ulster, 3 miles Northeast of Dungannon. Area, 4,461 acres. Population, 4,106.

TULLAGHOBIGLY.—(or RYETULLAGHOBIGLY.) Is a maritime parish of County Donegal, Ulster, 6 miles Southwest of Dunfanaghy. Area, 68,609 acres. Population, 9,049. It includes several islands.

TULLAGHORTON.—Is a parish of County Tipperary, Munster, 2 miles Northeast of Clogheen. Area, 6,889 acres. Population, 2,198.

TULLAGHOUGHT.—Is a parish of County Kilkenny, Leinster, 2¾ miles South Southwest of Kilmaganny. Area, 4,602 acres. Population, 1,750.

TULLAMORE.—Is an inland town of County Kings, Leinster, on an affluent of the Clodagh river and on the Grand Canal, 50 miles West Southwest of Dublin. Population, 6,342. It is the principal shipping station on the Grand Canal. Charleville forest, adjacent, is the seat of Earl Charleville, who owns the town.

TULLAROAN.—Is a parish of County Kilkenny, Leinster, 5 miles South Southwest of Freshford. Area, 12,360 acres. Population, 3,490. Here are the remains of Courtstown Castle.

TULLOW.—Is a market town and parish of County Dublin, Leinster, on the Slaney river, 8 miles East Southeast of Carlow. Area of parish, 7,990 acres. Population, 4,478 ; of town, 3,097. It has a Castle of the twelfth Century.

TULLY.—There are several parishes by this name, as follows, viz : County Dublin, Leinster, 3 miles South of Kingstown. Area, 3,286 acres. Population, 1,207 ; (or TULLYFERNE,) County Donegal, Ulster. Area, 16,612 acres. Population, 6,141. It comprises part of the town of Ramelton ; (or COGLANSTOWN,) County Kildare, einster, 1 mile South Southeast of Kildare. Area, 5,154 acres. Population, 1,279.

TULLYCORBET.—Is a parish of County Monaghan, Ulster, 4 miles North of Ballybay. Area, 8,913 acres. Population, 5,096.

TULLYLISH.—Is a parish of County Down, Ulster, 5¼ miles South Southeast of Portadown. Area, 11,707 acres. Population, 12,660.

TULSK.—Is a borough and village of County Roscommon, Connaught, 11 miles Northwest of Roscommon. Population, 133.

TUMNA.—(or JOEMONIA.) Is a parish of County Roscommon, Connaught, 1 mile North of Carrick-on-Shannon. Area, 8,189 acres. Population, 4,180.

TUMORE.—(or TOOMOUR.) Is a parish of County Sligo, Connaught, 3¼ miles Southeast of Ballymote. Area, 10,835 acres. Population, 3,319.

TURLOUGH.—Is a parish of County Mayo, Connaught, 3¾ miles Northeast of Castlebar. Area, 24,567 acres. Population, 7,430. In the village are the remains of an Abbey, and a very perfect Pillar-Tower.

TYBOHINE.—(or TAUGHBOYNE.) Is a parish of County Roscommon, Connaught, 4½ miles Northwest of Castlereagh. Area, 4,492 acres. Population, 17,804. It comprises the town of Frenchpark.

TYNAGH.—Is a parish of County Galway, Connaught, 7½ miles Northwest of Portumna. Area, 12,520 acres. Population, 5,941; of village, 348. The remains of a Church and four Castles are here.

TYNAN.—Is a parish of County Armagh, Ulster, 6½ miles West Southwest of Armagh. Area, 17,046 acres. Population, 11,392; of village, 177.

TYNISH.—Is an islet off the West coast of County Galway, Connaught, 2 miles West of Lettermore. Population, 150; mostly employed in fishing.

TYRONE.—Is an inland County of Ulster; having County Londonderry on the North, County Armagh on the East, County Monaghan on the South and County Donegal on the West. Area, 806,640 acres; of which 311,867 acres are uncultivated. Population in 1851, 254,878. Surface hilly; rising into mountains in the North and South. Soil is fertile; and well watered by the tributaries of the Blackwater and Foyle rivers. Agriculture generally very backward; the principal crops being Potatoes, Oats, Barley, Flax and Clover. The manufacture of Linen, Coarse Woolens, Blankets. Whiskey, Beer, Flour, Meal and Coarse Earthenware is carried on quite extensively. This County is subdivided into 4 baronies and 35 parishes in the dioceses of Clogher, Armagh and Derry. Strabane, Dungannon, Cookstown and Omagh are the principal towns. Tyrone sends two members to the House of Commons. The O'Neills, Kings in Ireland antecedent to Christianity, were chiefs of Tyrone when the memorable Rebellion of 1597 broke out under their auspices.

ULLARD.—Is a parish of County Carlow, Leinster, 2½ miles North Northeast of Graigue. Area, 5,848 acres. Population, 2,354. It has the remains of a Castle and some Ecclesiastical edifices.

ULLID.—Is a parish of County Kilkenny. Leinster, 2½ miles Northeast of Mountcoin. Population, 646.

ULSTER.—Is the most Northern Province of Ireland; lying between Latitude, 53° 46' and 55° 26' North. Longitude, 5° 24' and 8° 45' West. Having the province of Leinster on the South, Connaught on the Southwest, the Atlantic Ocean on the West and North and the Irish Sea on the East. Area, 5,475,438 acres; of which 1,764,370 acres are uncultivated, and 241,856 acres are under water. Population, 2,286,622; the majority of whom are Protestants. The shores are mostly bold and rocky, and on the North and East are remarkable basaltic cliffs, including the Gaints Causeway. The principal headlands are, Malin, Teelin and Fair Heads; and inlets, Donegal bay on the West, Loughs Swilly and Foyle on the North, and Belfast Lough and Dundrum bay on the East. Surface diversified and mountainous in the West, several summits rising to the height of 2,000 feet. The large lakes called Loughs, Neagh, Strangford and Erne, and the Bann, Foyle and Erne rivers, and some tributaries of the Shannon river, are in this province, and it is traversed by the Ulster Canal, 24 miles in length connecting Loughs Neagh and Erne. This province is the headquarters of the Irish Linen manufacture, and of other branches of manufacturing industry having their chief seat at Belfast. The annual value of Linens made is estimated at £5,000,000 and the manufacturers employ upward of 300,000 hands. The province is divided into the Counties, Donegal, Londonderry, Antrim, Down, Armagh, Cavan, Monaghan, Tyrone and Fermanagh.

URLINGFORD.—Is a market town and parish of County Kilkenny, Leinster, 15 miles West Northwest of Kilkenny. Area, 3,498 acres. Population, 2,830; of town, 1,742. It is neatly built. It has two Schools, a Bridewell and ruins of an old Castle.

URNEY.—Is a parish of County Tyrone, Ulster. Area, 14,489 acres. Population, 7,662. It comprises a small part of the town of Strabane. It is also the name of a parish of County Cavan, Ulster. Area, 7,935 acres. Population, 6,454. It comprises the town of Cavan.

VENTRY.—Is a maritime parish of County Kerry, Munster, 4 miles West Southwest of Dingle. Area, 4,439 acres. Population, 2,426. The harbor of Ventry, which affords good anchorage, is divided from that of Dingle by a narrow isthmus, on which are several Danish Entrenchments, said to have been the last Danish Military Post in Ireland.

VILLIERSTOWN.—Is a chapelry and village of County Waterford, Munster, 9½ miles North of Youghal. Population, 328.

VINEGAR-HILL.—Is in County Wexford, Leinster, East of Enniscorthy. It was, in 1798, the headquarters and scene of many of the atrocities of the Irish Insurgent forces.

VIRGINIA.—Is a small market town of County Cavan, Ulster, 15 miles Southeast of Cavan.

WARINGSTOWN.—Is a market town of County Down, Ulster, 6½ miles East of Portadown. Population, 825. It has large Cambric and Linen Manufactories.

WARRENPOINT.—Is a market town and parish of County Down, Ulster, at the mouth of the Newry river in Lough Carlingford, 6 miles Southeast of Newry. Area of parish, 1,178 acres. Population, 2,045; of town, 1,540. It is clean and well built. It is much resorted to for sea-bathing.

WATERFORD.—Is a maritime County of Munster; having Waterford harbor on the East, County Cork on the West, and Counties, Tipperary and Kilkenny on the North, from which it is separated by the Suir river. Area, 461,553 acres; of which 105,496 acres are waste. Population in 1851, 164,051. Surface is generally mountainous. The Knockmeledown and Cummeragh mountains intersect the County from West to East, and rise in some places to the height of 2,600 feet. The land is level and fine along the Suir river in the North and East, and the Blackwater and Bride rivers in the West, and skirting the coast, which is indented by Tramore, Dungarvon and Youghal harbors. Agriculture is improving. Average rent of land, 12s. 6d. per acre. The Fisheries are important. It has a few small manufactories. The County is divided into 7 baronies and 82 parishes in the dioceses of Waterford and Lismore, which, with Dungarvon, Portlaw, Tallow and Cappoquin, are the principal towns. The County sends two members to the House of Commons.

WATERFORD.—Is a city and County, parliamentary borough and seaport of County Waterford, Munster, on the Suir river, 29 miles South Southeast of Kilkenny and 85 miles South Southwest of Dublin. Area of borough, 10,059 acres. Population 29,283; of whom 24,783 are in the city. It is generally poorly built but has some handsome streets. Its quay and harbor are the finest in Ireland. It contains many handsome edifices, among which are, the Cathedral, Bishops Palace, Roman Catholic Cathedral, 5 other Churches, College of St. John, Town Hall, County and City Prisons, St. Reginald's Tower on the quay, an Ancient Fortress, (now a Police Barrack,) and several Hospitals, Schools, etc. Vessels of 800 tons can load and unload at quay. Waterford is the entrepot for a large extent of country, the exports of which are valued at two millions sterling annually. Markets, four times weekly. This city sends two members to the House of Commons and gives the title of Marquis to the head of the Beresford family, whose magnificent seat, Curraghmore, is in the vicinity, the demesne comprising 4,600 acres, traversed by the Clyde, and finely wooded.

WATERGRASS-HILL.—Is a market town of County Cork, Munster, 10 miles Northeast of Cork. Population, 801.

WATERSIDE.—Is a small town of County Londonderrry, Ulster, on the Foyle river. Population, 666.

WERBURGH.—(St.) Is a parish of County Dublin, Leinster. Area, 17 acres. Population, 2,969. It is comprised in the city of Dublin.

WESTMEATH.—Is an inland County of Leinster; having as a boundary, the Counties, Longford, Meath, Kings and Roscommon. Area, 453,468 acres; of which 56,392 acres are uncultivated. Population in 1851, 111,409. Surface undulating, diversified with woods, loughs and bogs. Soil fertile and well watered by the Shannon, Inny and Brosna rivers, and Loughs, Dereveragh, Ennel, Owhel, Lane, Iron, Sheelin, etc. Agriculture is increasing, the principal crops being Potatoes, Oats and Wheat. The Royal

Canal intersects the County, and a branch of the Grand Canal runs to Kilbeggan. This County is subdivided into 12 baronies and 63 parishes, chiefly in the diocese of Meath. The principal towns are Mullingar, Moat and a part of Athlone. It sends two members to the House of Commons. It gives the title of Marquis to the Nugent family.

WESTPORT.—Is a seaport town of County Mayo, Connaught, near Clew bay, 10 miles Southwest of Castlebar. Population, 4,365. It is one of the neatest towns in Ireland, and was well laid out by the first Marquis of Sligo. In its centre is a handsome space termed the Mall, from which the principal streets diverge at right angles. It has a parish Church, large Roman Catholic Chapel, Linen Hall, Court and Market Houses, etc. The Linen trade is large and it has an active export-trade. It gives the title of Viscount to the Marquis of Sligo, whose beautiful domain adjoins the town on the West; and besides which, the Reek, a mountain celebrated in the legendary history of Ireland, is in its vicinity.

WESTPORT QUAY.—Its port, 1 mile West, is at the Southeast extremity of Clew Bay termed Westport bay. Population, 547. It has a fishery and several coast guard stations.

WEXFORD.—Is a maritime County of Leinster; having the Irsh Sea and St. George's channel on the East and South, and on the other sides the Counties, Kilkenny, Carlow, Wicklow and Waterford. Area, 576,558 acres; of which 45,500 acres are uncultivated. Population in 1851, 180,154. Surface is mountainous in the Northwest, and declines to a level plain along the coast. Soil generally fertile, being watered by the Slaney river. The barony of Forth in the Southwest is occupied by decendants of a Welsh colony and is peculiarly well cultivated. Fisheries important. The principal towns are, Wexford, Enniscorthy, New Ross, Gorey and Newtownbarry. The County is subdivided into 9 baronies and 144 parishes in the dioceses of Ferns and Dublin. It sends two members to the House of Commons.

WEXFORD.—Is a parliamentary and municipal borough, seaport town and capital of County Wexford, Leinster, on the right bank of the Slaney river, where it expands into Wexford harbor, 12 miles South of Enniscorthy, and 64 miles Southwest of Dublin. Area of borough, 762 acres. Population in 1851, 12,471. It is poorly built, but the Quay and one or two other streets are lined with good buildings. There are some remains of ancient walls, of an Abbey, and other Ecclesiastical edifices ; outside of the town is a fine granite Column, in memory of the exploits in Egypt by the army under Abercrombie. Wexford has Protestant, Diocesan and other Schools, Chamber of Commerce, etc. It sends one member to the House of Commons.

WHIDDY ISLAND.—Is an island off the coast of County Cork, Munster, in Bantry bay, near its head, 3 miles in length. On it are a coast-guard station and several Forts for the defense of Bantry harbor. Population, 450.

WHITECHURCH.—There numerous parishes by this name, as follows, viz : County Waterford, Munster, 5 miles West Northwest of Dungarvon. Area, 9,952 acres. Population, 3,403 ; County Cork, Munster, 5½ miles North Northwest of Cork. Area, 10,515 acres. Population, 3,368 ; or (GLYNN,) County Wexford, Leinster, 2½ miles Northeast of Taghmon. Area, 7,188 acres. Population, 1,960 ; County Wexford, Leinster, 5 miles Southwest of New Ross. Area, 5,342 acres. Population, 1,384 ; County Dublin, Leinster, 1½ miles South of Rathfarnham. Area, 2,873 acres. Population. 13,075 ; County Tipperary, Munster, 3 miles Southwest of Cahir. Area, 3,922 acres. Population, 1,274 ; County Kilkenny, Leinster, 2 miles Northwest of Pilltown. Area, 2.187 acres. Population, 837 ; County Kildare, Leinster, 2¼ miles North of Kill. Area, 3,166 acres. Population, 320.

WICKLOW.—Is a maritime County of Leinster ; having the Irish sea on the East, and the Counties, Dublin, Kildare, Carlow and Wexford on the North, West and South. Area, 500,178 acres ; of which 200,745 acres are uncultivated. Population in 1851, 98,978. Surface is precipitous. Soil fertile in the lowland, and watered by the Liffey and Slaney rivers in the West, and the Ovoca and Vartrey rivers in the East, all of which rise in the County. The principal crops are Oats and Potatoes. From 10,000 to 12,000 tons of Copper Ore, and from 1,400 to 3,800 tons of Lead are produced Annually ; large quantities of Sulphuret of Iron and some Gold are met with. The manufactories have declined. Its principal seat is Stratford on-Slaney. Wicklow, Arklow and Bray are the principal towns. This County sends one member to the House of Commons. At Glandalagh, or Glandalough, formerly an Episcopal See in this County, is one of the finest collection of ruins in the United Kingdom, termed the Seven Churches.

WICKLOW.—Is a seaport town and capital of County Wicklow,

Leinster, at the mouth of the Vartrey, 27 miles Southeast of Dublin. Population, 2,794. It is resorted to for sea-bathing. It has some important trade, and exports Copper Ore and Corn. The harbor is shallow. Races are held here Annually. It gives the titles of Earl and Viscount to the Howard family.

WICKLOW HEAD.—Is about 2½ miles East Southeast of Wicklow, and is surmounted by two Light-Houses, respectively 540 and 250 feet in height. Latitude, 52° 57′ 9″ North. Longitude, 6° West.

YOUGHAL.—Is a parliamentary and municipal borough, seaport town and parish of County Cork, Munster, 27 miles East of Cork, on the West side of the estuary of the Blackwater river, which forms its harbor. Area of parish, 4,830 acres. Population, 12,034. Area of parliamentary borough, 344 acres. Population in 1851, 7,410. It stands at the foot of a steep height, and was formerly enclosed by walls flanked with towers; parts of which remain and outside of which are some poor suburbs. The town is antiquated and its main street is crossed by an old archway, beside which its chief structures are the large Gothic parish Church, (containing the tomb of the great Earl of Cork and near it the ruins of an Abbey,) Chapel of Ease, Roman Catholic and other Chapels, Town House, Court House, Custom House, etc.; and the House of Sir Walter Raleigh, which is preserved nearly entire. It has an active trade in rural produce. It has some Potteries, Brickworks, etc., and a valuable Salmon Fishery. It sends one member to the House of Commons. This is believed to be the spot where Sir Walter Raleigh first introduced the culture of the potatoe into Ireland.

YOUGHLARRA.—Is a parish of County Tipperary, Munster, 5 miles West Northwest of Nenagh. Area, 8,356 acres. Population, 3,321. It contains a hamlet named Youghal, and the remains of several Feudal and Ecclesiastical edifices.

ROUND TOWERS IN IRELAND.

Name.	County.	Name.	County.
Aghadoe,	Kerry.	Kilcullen,	Kildare.
Aghagower,	Mayo.	Kildare,	Kildare.
Antrim,	Antrim.	Kilkenny,	Kilkenny.
Ardfert,	Kerry.	Killala,	Mayo.
Ardmore,	Waterford.	Kilmacduagh,	Galway.
Ballagh,	Mayo.	Kineth,	Cork.
Ball,	Sligo.	Kilree,	Kilkenny.
Ballygaddy,	Galway.	Limerick,	Limerick.
Boyle,	Roscommon.	Lusk,	Dublin.
Brigoon,	Cork.	Mahera,	Down.
Ballynerk,	Cork.	Melic,	Galway.
Cailtre-Isle,	Clare.	St. Michael,	Dublin.
Cashel,	Tipperary.	Moat,	Sligo.
Castledermot,	Kildare.	Monasterboice,	Louth.
Clondalkin,	Dublin.	Newcastle,	Mayo.
Clones,	Monaghan.	Nohovel,	Cork.
Clonmacnois,	Westmeath.	Oran,	Roscommon.
Cloyne,	Cork.	Oughterard,	Kildare.
Cork,	Cork.	Ram Isle,	Antrim.
Devenish,	Fermanagh.	Rathmichael,	Dublin.
Donoghmore,	Meath.	Rattoo,	Kerry.
Downpatrick,	Down.	Roscrea, two,	Tipperary.
Drumboe,	Down.	Scattery,	Clare.
Drumcliff,	Sligo.	Sligo, two,	Sligo.
Drumiskin,	Louth.	Swords,	Dublin.
Drumlahan,	Cavan.	Teghadon,	Kildare.
Dyfart,	Queens.	Timahoe,	Queens.
Ferbene, two,	Kings.	Tulloherin,	Kilkenny.
Fertagh,	Kilkenny.	Turlogh,	Mayo.
Glondaloch, two,	Wicklow.	West Carbury,	Cork.
Kilberman,	Galway.		

DESCRIPTION OF PRINCIPAL

ROUND TOWERS IN IRELAND.

Name.	Height.	Circum-ference.	Thickness of Walls.		Door from the Ground.	
	Feet.	Feet.	Feet.	Inches.	Feet.	Inch.
Cloyne,	92	50	3	8	13	..
Fertagh,	112	48	3	8	10	..
Kilcullen,	40	44	3	6	7	..
Kilmacduagh,	110	57	2	7	24	..
Teghadow,	71	38	3	8	11	6
Downpatrick,	66	47	3	..	.	..
Devenish,	76	41	3	6	..	..
Monasterboice,	110	51	3	6	6	..
Timahoe,	35	53	4	4	14	..
Kildare,	110	54	3	6	13	..
Oughterard,	25	48	3	..	8	..
Cashel,	55	54	4	..	11	..
Swords,	95	55	4	8	2	..
Abernethy,	57	49	3	..	3	..
Brechin,	85	51	..	..	..	..
Druniskin,	130	..	..	..	..	..
Kenith,	70	..	..	..	..	..
Kells,	99	..	..	..	..	..

It is remarked that almost all the Round Towers are divided into stories of different heights, the floors supported in some by projecting stones, in others by joists put in the wall at building, and in many they were placed upon rests; the latter are from four to six inches, carried round and taken off the thickness of the wall in the story above. And I conjecture these rests diminish the thickness of the wall. Although some of our writers on Ireland say that a great number of these towers are thicker at the top than at the base. Some of our towers seem thicker at top, but that is from a swelling in the walls, which rather adds to the thickness.

Cashel Tower is divided into five stories, with holes for joists; Tertagh has five stories and one rest; Kilcullen has three stories and one rest; Kildare has six stories, and projecting stones for each; Oughterard has five stories, and projecting stones; Teghadow has six stories; Monasterboice has six stories and projecting stones; Timahoe has five stories. The door of Cashel Tower faces Southeast; those of Kilkenny and Kildore South, and the others vary. Kenith Tower stands 124 feet high; Drumboe 20 feet high; Downpatrick 48 feet high; Kildare 90 feet high; Monaghan Tower is 60 feet high and 15 feet in diameter; the door 5 feet high by 2 feet wide; Mayo Tower is 84 feet high, 51 feet in circumference, the door plain, and 5½ feet high by 2½ feet wide. This is the most prominent List of the Round Towers in Ireland, and the most worthy of notice.

www.ingramcontent.com/pod-product-compliance
Lightning Source LLC
Chambersburg PA
CBHW020248170426
43202CB00008B/277